HUMAN VALUES

AND

Science, Art and Mathematics

Books with text by Lillian R. Lieber
and drawings by Hugh Gray Lieber

The Education of T. C. Mits
(W. W. NORTON & COMPANY, INC.)

The Einstein Theory of Relativity
(HOLT, RINEHART & WINSTON)

Take a Number
(RONALD PRESS COMPANY)

Galois and the Theory of Groups
(GALOIS INSTITUTE PRESS)

Non-Euclidean Geometry
(GALOIS INSTITUTE PRESS)

Good-bye Mr. Man, Hello Mr. NEWman
(GALOIS INSTITUTE PRESS)

Mits, Wits and Logic
(W. W. NORTON & COMPANY, INC.)

Infinity
(HOLT, RINEHART & WINSTON)

Comedie Internationale
(GALOIS INSTITUTE PRESS)

Lattice Theory: The Atomic Age in Mathematics
(GALOIS INSTITUTE PRESS)

Human Values and Science, Art and Mathematics
(W. W. NORTON & COMPANY, INC.)

HUMAN VALUES and SCIENCE, ART AND MATHEMATICS

Text by
LILLIAN R. LIEBER

Drawings by
HUGH G. LIEBER

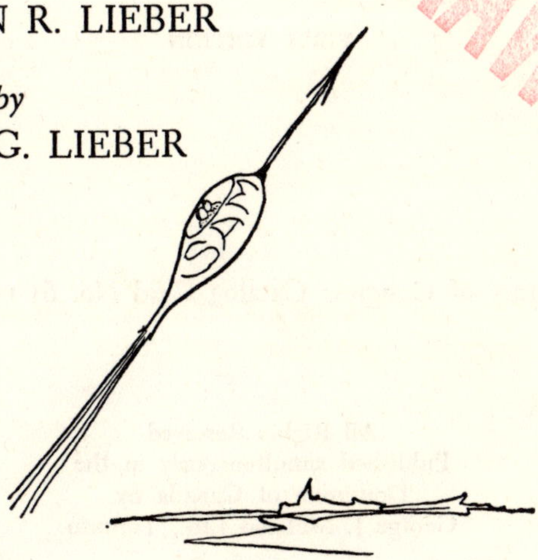

NEW YORK
W. W. Norton & Company, Inc.

FIRST EDITION

Library of Congress Catalog Card No. 61-11342

Published simultaneously in the
Dominion of Canada by
George J. McLeod Ltd., Toronto

PRINTED IN THE UNITED STATES OF AMERICA
FOR THE PUBLISHERS BY THE VAIL-BALLOU PRESS

1 2 3 4 5 6 7 8 9

DEDICATION

This book is
affectionately dedicated to
Tristram Walker Metcalfe
who was
a distinguished educator,
in the best sense of the term—
he brought out in all people
what was best in them
and then gave them
the freedom to grow,
whether it be in
Art or Mathematics
or any other
fine human enterprise.
He believed in the human race
and was determined
to help all people to realize
the best that was in them
despite the
"Killers of the Dream."
We are in desperate need
of more of such
courageous,
wise,
dedicated educators
TODAY!

This is not intended to be
free verse.
Writing each phrase on a separate line
facilitates rapid reading,
and everyone
is in a hurry
nowadays.

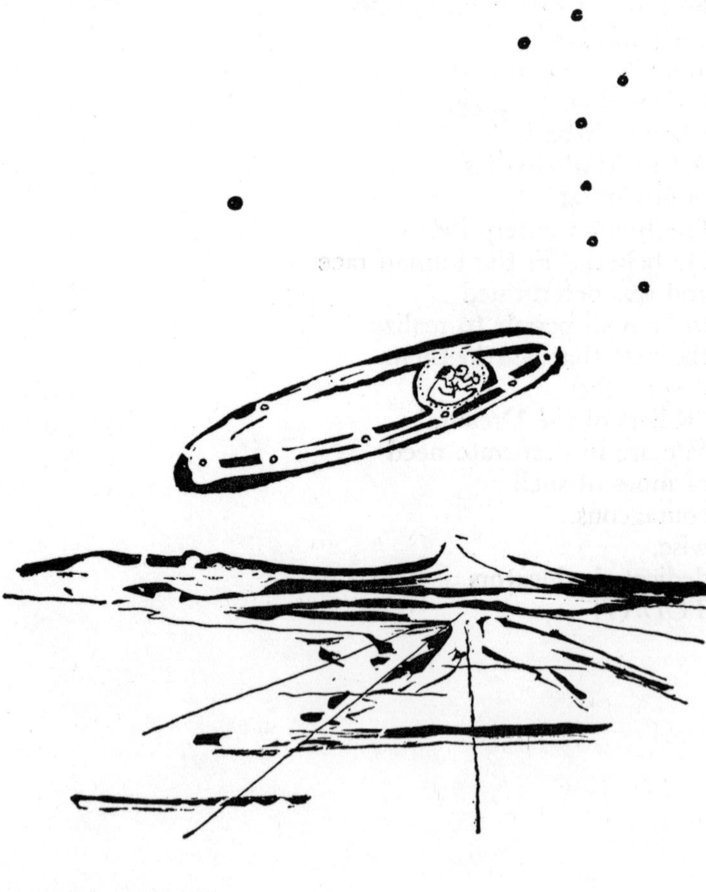

PROLOGUE AND EPILOGUE

This should be read
both before and after the rest of the book—
before, in order to get the flavor of things to come;
after, to judge whether the promises have been kept.

This book is really about
Life, Liberty, and the Pursuit of Happiness,
using ideas from mathematics
to make these concepts less vague.
We shall see first what is meant by
"thinking" in mathematics,
and the light that it sheds on both the
CAPABILITIES and the LIMITATIONS
of the human mind.
And we shall then see what bearing this can have
on "thinking" in general—
even, for example, about such matters as
Life, Liberty, and the Pursuit of Happiness!

For we must admit that our "thinking"
about such things,
without this aid,
often leads to much confusion—
mistaking LICENSE for LIBERTY,
often resulting in juvenile delinquency;
mistaking MONEY for HAPPINESS,
often resulting in adult delinquency;
mistaking for LIFE itself
just a sordid struggle for mere existence!

Let me indicate how mathematics can help us—
for example by showing what happened in mathematics
at the beginning of the 19th century,
which led to a great

NEW FREEDOM that was NOT LICENSE!
By such specific illustrations,
issues can be stated more clearly
and considered more easily,
without the usual confusion.
Thus let us take a brief look at
the history of mathematics as depicted in
the single-line diagram at the right
(drawn approximately to scale)!
These are significant dates because
so far as we know

(1) the earliest mathematics goes back to about 5000 B. C.,*
(2) the first WRITTEN document on mathematics, the well-known Ahmes papyrus,* dates back to about 1800 B. C.,
(3) the golden period of Greek * mathematics (the work of Euclid and others) belongs to the time around 300 B. C.,
(4) the 19th century, 1800 to 1900 A. D.,* may be designated as the beginning of MODERN mathematics.

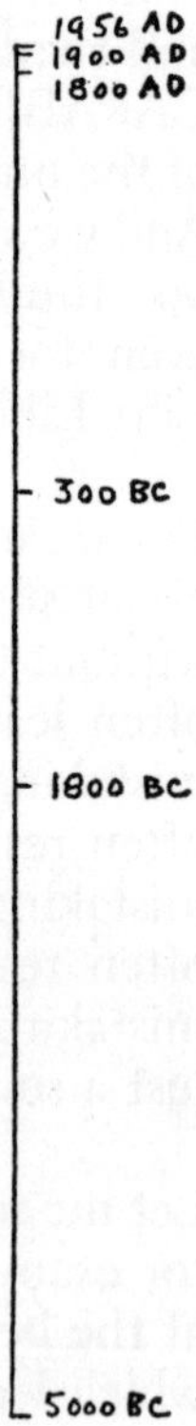

Now it is a fact that
the amount of mathematics created
during the 19th century alone
was more than FIVE TIMES as much as
was created in all of previous history! *

WHY?!

You will see in this little book
that at the beginning of the 19th century
man became aware, IN MATHEMATICS,
how much more CREATIVE he can be!
And this NEW FREEDOM opened up

* See E. T. Bell's wonderful book: *The Development of Mathematics.*

the floodgates of discovery,
as mentioned above.
And so,
by looking at what happened in mathematics,
we can come to a better idea of
what MAN is capable of,
and HOW he can utilize his capabilities.
Moreover, by looking at mathematics
one can also see that
in spite of this great discovery,
namely, the realization that
postulates are MAN-MADE *—
still this does NOT mean that
man does not need GOD (!),
and does NOT lead to human ARROGANCE,
but rather to HUMILITY!—
for we find in mathematics
specific instances of the idea that
the MORE we KNOW, the MORE we contact also
the UNKNOWN.
This may sound paradoxical,
but consider the following diagram

in which the finite region of the "KNOWN"
is surrounded by the
UNBOUNDED region of the "UNKNOWN";
then, as we enlarge the "KNOWN" region,
note that the BOUNDARY
between the KNOWN and the UNKNOWN

* See Bell's first chapter, "Freedom and Responsibility."

INCREASES,
(shown by the dotted line in the diagram),
as witness the following example:
There is a new branch of mathematics
known as LATTICE THEORY,
developed largely by
the great, young, brilliant American mathematician,
Garrett Birkhoff,
at Harvard University.
He published a book about it in 1940,
which is a magnificent treatise on
this powerful new mathematical tool;
also in it he listed 17 problems in this domain
still unsolved at that time.
Later, in 1948, his second edition of this book
had to be almost completely re-written
because of the tremendous advances
which had been made in this domain
during this short interval;
and he now stated that
8 of the 17 problems had already been
essentially solved,
BUT
he now listed 111 NEW UNSOLVED problems!

Does this discourage you?
Or do you react to it with the
"cheerful resignation"
which a mathematician learns from
the realization of the
LIMITATIONS of his great subject
WITHOUT losing his deep respect for its
POSITIVE ACCOMPLISHMENTS,
thus giving to MAN
his dignified but NOT OMNISCIENT place
in the Universe.

These are comforting thoughts,

and mathematics can teach us to accept them!
It is for this reason that
EVERYONE should know something of
this approach to this beautiful subject,
and not regard it as just a useful means of
counting your small change!

PREFACE

This little book
is intended to show that
"Science" and "The Humanities"
are NOT two separate domains
at war with each other,
but, on the contrary,
are one and the same!
Nowhere can one see this
so EASILY and CLEARLY
as in
even a brief look at
MODERN MATHEMATICS and SCIENCE and
MODERN ART,
which show the
INFINITE VARIETY that
MAN is capable of,
how great is his marvelous
creative ability,
if only he would take
a good look at himself
and give up the false notions
that have been fed to him
that he is just a
greedy
violent
MONSTER!

CONTENTS

HUMAN VALUES
AND
Science, Art and Mathematics

I. FREEDOM AND RESPONSIBILITY

1. The Great Discovery of 1826.

One of the amazing things
in the history of mathematics
happened at the beginning of
the 19th century.
As a result of it,
the floodgates of discovery
were opened wide,
and the flow of creative contributions
is still on the increase!
As E. T. Bell,*
the distinguished American mathematician,
says:
"The nineteenth century . . .
contributed to mathematical knowledge
about FIVE TIMES AS MUCH as was done
in the whole of preceding history."

And the rate of increase
has been even greater
since 1900!

Furthermore,
this amazing phenomenon
was due to a mere
CHANGE OF ATTITUDE!
Perhaps I should not say "mere,"
since the effect was so immense—
which only goes to show that
a CHANGE OF ATTITUDE
can be extremely significant,

* E. T. Bell: *Men of Mathematics*, p. 17.

and we might do well
to examine our ATTITUDES
toward many things, and people,—
this might be most rewarding,
as it proved to be in
mathematics.

As we shall see.

2. A Change of Attitude.

In order to appreciate a
CHANGE of attitude,
we must of course know
what was the OLD attitude,
and what is the NEW one.
To do this,
we must go back to
Euclid, about 300 B. C.,
who,
as every high school boy knows,
first put together
the various known facts of geometry
into a SYSTEM,
instead of leaving them as
isolated bits of information—
as in a quiz program!

If you have ever had occasion
to use a system,
you probably know that by this means
one can learn more,
faster,
and even use it to
make new discoveries!
Are you skeptical when
you remember your struggles with
Euclidean geometry

in high school?
For perhaps you did NOT learn
much,
nor fast,
nor discover anything new
with its help!

Perhaps you are in the state of mind
of some art students
who were being introduced to
Birkhoff's theory of aesthetic measure.*
The artist, Hugh Gray Lieber,
who was telling them about it,
wanted to draw some simple figures,
triangles, quadrilaterals, etc.—
in order to see which shapes
seemed to them to be
"more" or "less" pleasing,
and see if they agreed with
Birkhoff's way of
measuring
how pleasing a shape can be!
To the artist's amazement,
no sooner did he draw some triangles,
than a general groan went up
from the students—
it seems they were reminded of
their geometry in high school
which they had learned to dislike
so intensely,
that the very thought of
finding these "shapes"
even mildly pleasing
seemed to them preposterous!

But let us not blame this on Euclid,
nor on the idea of a SYSTEM—

* G. D. Birkhoff: *Aesthetic Measure.*

for people can even learn to
hate Shakespeare and Beethoven!—
the fault surely lies elsewhere.
This however is a subject
for another essay;
perhaps you will want to write this one
yourself.

The fact is that
Euclid's system
has served for many centuries
as a MODEL for clear thinking,
and has been and still is
of the greatest value to the human race.

Now, what is a "SYSTEM"?

In constructing a system,
one must begin with
a few simple statements
from which,
by means of logic,
one derives the "consequences."
We can thus
"figure out the consequences"
before they hit us.
And this we certainly need more of!

Thus Euclid started with such
simple statements
(called "postulates" in mathematics)
as:
  "It shall be possible to draw
  a straight line joining
  any two points,"
and others like it.*

* (1) T. L. Heath: *Thirteen Books of Euclid's Elements.*
(2) J. W. Young: *Fundamental Concepts of Algebra and Geometry.*

From these
he derived many complicated theorems
(the "consequences")
like the well-known
Pythagorean Theorem,
and many, many others.

And, as we all know,
to "prove" any theorem
one must show how
to "derive" it from the postulates—
that is,
every claim made in a "proof"
must be supported by reference to
the postulates or
to theorems which have previously
already been so "proved"
from the postulates.
Of course Theorem #1
must follow from
the postulates ONLY.

Now what about
the postulates themselves?
How can THEY be "proved"?
Obviously they
CANNOT be PROVED at all—
since there is nothing preceding them
from which to derive them!
This may seem disappointing to those who
thought that in
Mathematics
EVERYTHING is proved!
But you can see that
this is IMPOSSIBLE,
even in mathematics,
since EVERY SYSTEM must necessarily
START with POSTULATES,

and these are NOT provable,
since there is nothing preceding them
from which to derive them.

Of course Euclid was
well aware of this.
One might ask,
"Why, then, did Euclid consider
his system to be reliable
when he knew that
it was BASED on UNPROVED postulates?"
The answer is that
he considered the postulates to be
"SELF-EVIDENT TRUTHS."

And so
Euclid's system may be
represented diagrammatically thus:

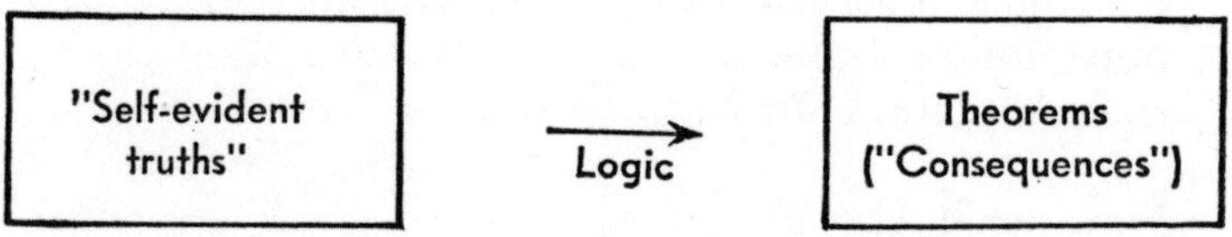

This then is
what I have referred to as
the "OLD" point of view,
which was basically changed
in the beginning of the 19th century,
leading to
the Great Discovery of 1826.

Now what is the "NEW" point of view,
and what brought about this
CHANGE?
It is interesting to note that
creative mathematicians
are really very CONSERVATIVE,

making progress
not by a wanton "radical" approach,
but by an inevitable pressure
which makes change IMPERATIVE.
This was indeed the case
in the above-mentioned
"OLD" viewpoint toward POSTULATES,
which held them to be
"self-evident truths,"
and you will soon see
how mathematicians were
actually FORCED
to make the change,
after centuries of
FAILURE to solve a certain problem
by other means!

3. The Failure Which Led To Success.

Among Euclid's postulates
was the one now known as
the FAMOUS PARALLEL POSTULATE.
It may be stated thus:
Through any point C which is
NOT on a given line AB,

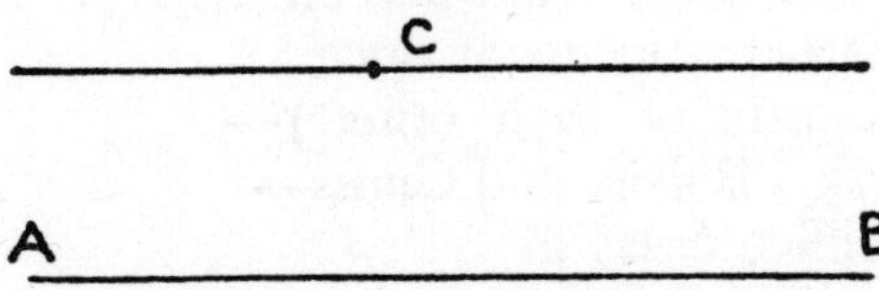

ONE and ONLY ONE line
can be drawn which is
PARALLEL to the given line—
that is,
which does NOT meet the given line,
no matter how far they are both prolonged
either to the right or to the left.

Now this particular postulate
did not seem to Euclid to be
a genuinely "self-evident truth";
and therefore
he felt that he should be able
to "prove" it
from his other postulates.
And he proceeded to try to prove it.
But he was UNABLE to do so!

Furthermore,
many other distinguished mathematicians,
after Euclid,
also tried to prove
the PARALLEL POSTULATE,
but also met with
FAILURE!
And this went on for centuries!

Even the great Legendre,*
as late as 1800,
gave what is now known to be
a fallacious "proof."

And then, in 1826, it dawned on
several mathematicians at once
(independently of each other!)—
Lobachevsky, Bolyai, and Gauss—
that the difficulty lay in
the very ATTITUDE toward
what postulates are—
that
instead of regarding them as

* You may be interested to see Legendre's argument in a paper on "The Euclidean Parallel Postulate" by Alonzo Church of Princeton University, published by The Galois Institute of Mathematics and Art, 624 A Third Street, Brooklyn 15, N. Y.

"self-evident truths,"
they should rather be regarded as
MAN-MADE ASSUMPTIONS.

Now why is this so IMPORTANT?
Does such philosophic hair-splitting by
"eggheads"
really matter?
One might think that,
whatever the NATURE of
the postulates themselves,
one should be able to say
whether or not
a statement
(in this case, the parallel postulate)
does or does not follow
from the postulates.

But NO!

For, if the statement is
a MAN-MADE ASSUMPTION,
(an "IFFY" statement,
as FDR used to say)
there is nothing SACRED about it,
it is NOT INEVITABLE,
and THEREFORE
may be CHANGED.
Now what has this to do with
the above-mentioned issue of
whether or not
the parallel postulate
follows from the OTHER Euclidean postulates?

You will soon see the connection.

The above-mentioned mathematicians
not only accepted this

NEW ATTITUDE toward
a postulate set,
but actually proceeded
TO CHANGE THE PARALLEL POSTULATE,
leaving the OTHER Euclidean postulates
UNCHANGED,
with the result that
they found
an entirely NEW GEOMETRY,
in many ways quite different from
Euclid's,
but just as SELF-CONSISTENT as
Euclid's.
Now, what does this mean?

You will see in the following sections
how these men changed
Euclid's parallel postulate,
what their NEW PARALLEL POSTULATE was,
and how their exhibiting
a NEW self-consistent geometry
shows that the parallel postulate is
INDEPENDENT of
the remaining postulates in the set,
that is,
it does NOT FOLLOW from them—
no wonder it could not be
"PROVED"!

You will also see that
although this new geometry
seemed "queer,"
it turned out to be
extremely useful—
not only in explaining
the dilemma of the parallel postulate,
but turned out to be even of
great PRACTICAL value as well!

Thus the NEW viewpoint
toward postulates
may be represented thus:

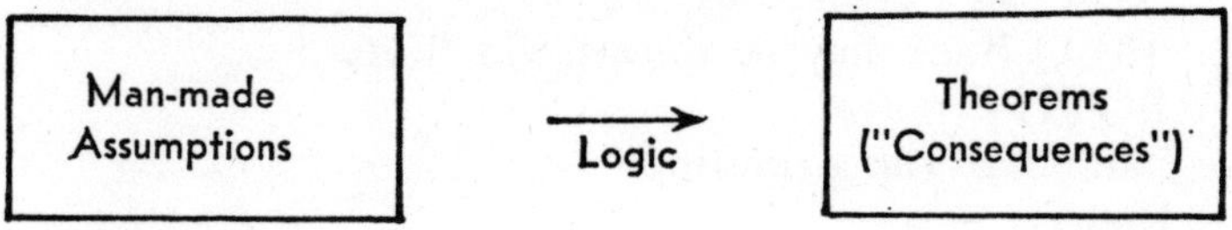

And this
SEEMINGLY SMALL CHANGE,
this "mere" change in ATTITUDE,
was the Great Discovery of 1826
which opened the floodgates of
creative work in mathematics
and science—
as you will soon see.

4. Non-Euclidean Geometries.

Lobachevsky, Bolyai, and Gauss
who, as I mentioned on page 26,
worked entirely independently of each other,
not only all got
the same brilliant idea about
the NATURE of the postulates themselves,
but they all decided to
CHANGE
ONLY the PARALLEL POSTULATE,
leaving the OTHER Euclidean postulates
as they were,
and, furthermore,
they all decided to change it in
the SAME way: *

* Whether this is "mind-reading"
is of course a most interesting question,
but, as of now,
nobody knows the answer,
and it will probably be investigated

namely,
to make it read:
Through any point C which is
NOT on a given line AB,
TWO lines may be drawn which are
BOTH
parallel to the given line—
that is,
which do NOT meet the given line,
no matter how far they are prolonged
either to the right or to the left.
Now, if you try to draw
a diagram like this

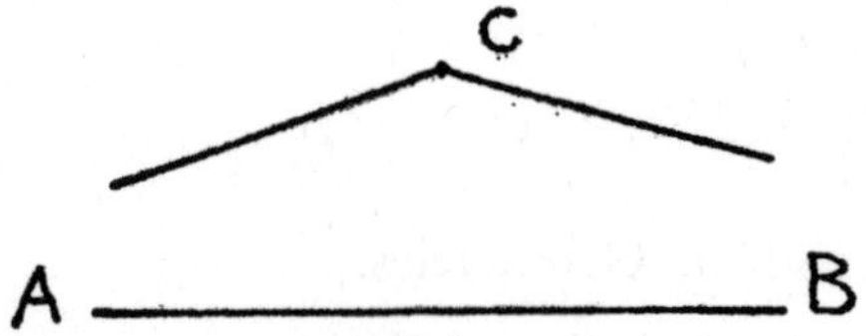

you will of course feel that
NEITHER line through C
can possibly be parallel to AB,
since either one, if prolonged,
will surely meet AB somewhere
(or AB prolonged).
But the fault is not
with the assumption that
two such lines are possible,
BOTH parallel to AB,
but rather with this misleading diagram—

---

in the future—we hope!
All we know now is that
this sort of thing has happened many times.
Compare the case of the calculus,
developed independently by
both Newton and Leibnitz,
as well as many other cases!

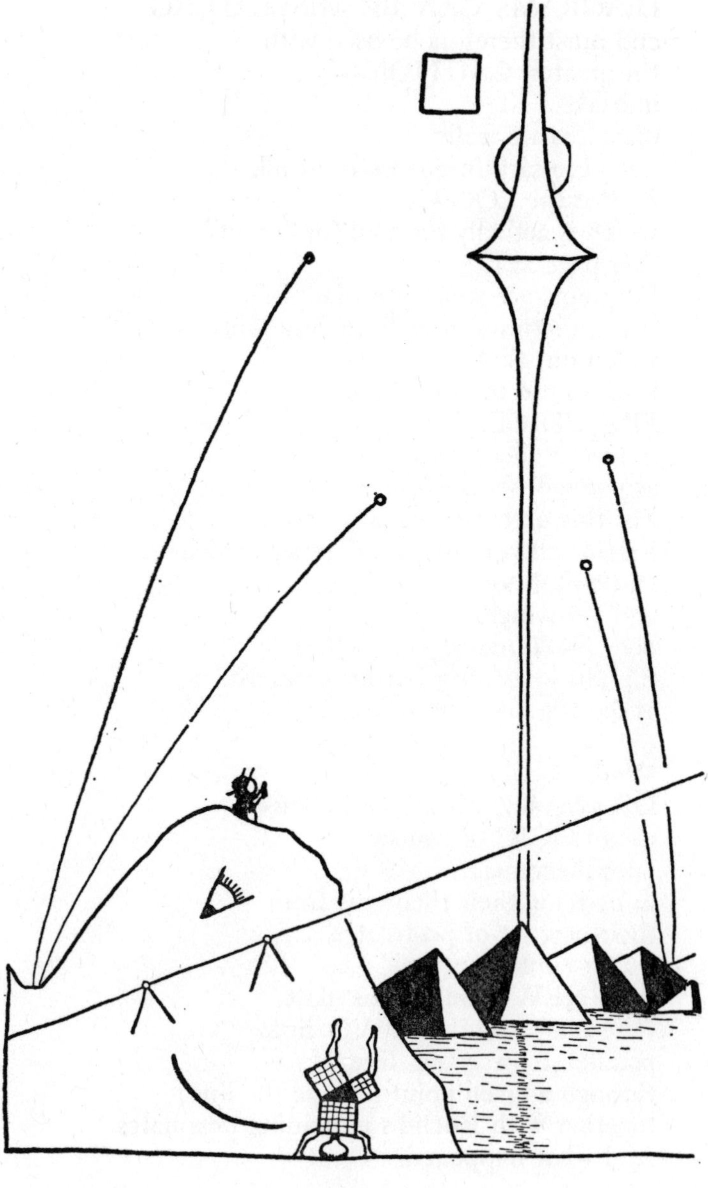

Let this be a WARNING that
DIAGRAMS CAN BE MISLEADING
and must therefore be used with
the greatest CAUTION—
indeed,
diagrams are really
not necessary in geometry at all,
for it is the LOGIC
which is actually the tool for "proof"
(see page 22).
Diagrams are sometimes helpful,
but sometimes actually a hindrance—
watch out that
they do not involve some
PREJUDICE,
as in the above case,
as you will soon see.
For this particular case,
I shall tell you about a diagram which
IS helpful here,
and yet which
does NOT make
the above assumption look absurd,
as did the diagram on page 30.

Well,
Lobachevsky, Bolyai, and Gauss
used LOGIC of course
(not diagrams)
in deriving their theorems from
their new set of postulates,
which consists of
their NEW parallel postulate
(the one assuming TWO lines
parallel to a given line
through a given point outside the line)
together with Euclid's remaining postulates.
And what happened?

They obtained many theorems
which seemed very bizarre—
for example:
(1) Two parallel lines are
NOT everywhere equally distant,
as they are in Euclidean geometry.
(2) The angle-sum of a triangle
is LESS than 180°,
(instead of exactly 180° as in
Euclidean geometry),
and is NOT always the same,
regardless of the
size and shape of the triangle
(as in Euclidean geometry),
but ranges from
0° to 180°;
etc., etc.
Thus an entirely different geometry
was created.
As Bolyai wrote to his father—
he had created a whole new world
out of his own head! *

Perhaps this only makes you
suspicious of the whole procedure!
Making you believe that
all this is merely
a mental exercise,
serving no real purpose.
You will soon see, however,
that
not only does it correspond to reality,
but that we actually live in a
NON-Euclidean world!

But first
let us see what happened next.

* H. S. M. Coxeter: *Non-Euclidean Geometry*, p. 10.

The realization that
postulates are
only man-made assumptions
led, within a few years,
in 1850,
to still another geometry:
this time, Riemann,
again leaving all the Euclidean postulates
unchanged, EXCEPT ONLY
the PARALLEL POSTULATE,
which he changed to:
Through any point C which is
not on a given line AB,
NO line can be drawn which is
parallel to the given line.

And again
this new postulate set
(and it IS a NEW SET even though
only a single postulate has been changed!)
led to still another
Non-Euclidean geometry,
different from the previous one
as well as from
our old Euclidean geometry,
again with some
seemingly bizarre theorems
like:
(1) From a given point outside a given line
many lines can be drawn which are
all perpendicular to the given line
(instead of only one as in
Euclidean geometry).
(2) The angle-sum of a triangle
is GREATER than 180°,
ranging from 180° to 540°.
And many, many others.

Again this SEEMS to make no sense—
and indeed it seemed to make no sense
even to the contemporaries of
these great men, who were, in consequence,
ignored by them! *

So you may wonder
why these queer geometries
are considered to be
significant and important now!—
aside from the fact that
the very existence of these
self-consistent non-Euclidean geometries
shows that
the parallel postulate is obviously
INDEPENDENT of
the remaining Euclidean postulates,
for, if it followed from them,
it could not be replaced by
its contradictory—
with either TWO or NO parallels through C—
as did the above-mentioned mathematicians!
No wonder their predecessors
were unable to prove it!

But let us now see other amazing uses
of the
Non-Euclidean Geometries!

5. What is the Use of It?

Well, in the first place,
shortly after all this happened,
namely in 1868,
a mathematician named Beltrami

* See the thrilling stories in
Men of Mathematics by E. T. Bell.

looked at these
Non-Euclidean geometries
in quite a different way!
He pointed out that
whereas these geometries
seem queer when applied to
the ordinary Euclidean "flat" surface,
they make perfectly good sense on
other, CURVED, surfaces.

For example,
consider the Non-Euclidean geometry of
Riemann
(see page 34)—
you will see that
many of its theorems,
like for instance (1) and (2) on page 34)
apply on the surface of a
SPHERE,
though NOT on a "flat" surface.
But how is this possible when
one cannot even draw a straight line
on a sphere?
Well,
there is a property of a straight line
which CAN apply on a sphere—
namely, the property that it is
THE SHORTEST DISTANCE BETWEEN TWO POINTS.
Thus, on ANY surface,
one can of course draw any number of paths
from point A to point B, like this:

and many, many others;
of all these,
there is a SHORTEST.
Now on a "flat" surface,
the shortest path is
an ordinary straight line,
whereas on a sphere
it is an arc of a great circle.*
Similarly,
other surfaces have
their own characteristic
shortest distance paths,
called "geodesics"—
thus the geodesic on a "flat" surface
is a straight line,
the geodesic on a sphere is
an arc of a great circle,
etc., etc.

If we now take
the equator on a sphere
as the GIVEN "straight line"
(meaning "shortest distance path" or
geodesic,
as explained above),
and the "north pole" as
the point, C, outside the line,
it is quite easy to see that
any meridian through C
meets the equator
perpendicularly
(since it forms
two equal right angles

* A great circle is obtained by
"cutting" the sphere in half—
thus the equator or any meridian, etc.
is a great circle,
whereas the latitude lines are
circles, but not GREAT circles.

where they meet),
so that we have here
any number of perpendiculars
from a point to a line
(as stated in (1) on page 34),
and,
although this is not the case on
a flat surface,
it is nevertheless very sensible
if we consider instead
the surface of a sphere.

Similarly, (2) on page 34
is also very sensible on this surface,
for, consider a triangle
(whose sides are of course
arcs of great circles,
which are the geodesics here)
whose base is
a part of the equator
and whose other two sides are
parts of two meridians
extending from the ends of the base
and meeting at the pole.
Here the two base angles are
both right angles (each 90°),
whereas the third angle,
the one at the pole,
may have a varying number of degrees
(depending on the length of the base chosen)
so that
the angle-sum of the triangle
is obviously MORE than 180°,
as claimed on page 34.

And so on,
for other theorems
which seem impossible when

we are thinking of a flat surface,
but are seen to be most sensible when
applied to a different surface.

Thus, also, it is easy to see that
two great circles on a sphere
always intersect each other,
which means that
two "straight lines" (i.e. geodesics)
on a sphere are
NEVER PARALLEL,
and we thus have a surface on which
Riemann's PARALLEL POSTULATE
(see page 34)
actually applies,
and is NOT QUEER at all!
It only seemed queer so long as
we were prejudiced in favor of
drawing figures on a flat surface
and studying the corresponding
Euclidean geometry.

Thus the
NEW ATTITUDE TOWARD POSTULATES
(page 29)
has succeeded in freeing us from this
PREJUDICE,
resulting in making it possible
for us to think of other surfaces and
their own, Non-Euclidean, geometries.

Similarly,
the PARALLEL POSTULATE of
Lobachevsky, Bolyai, and Gauss
was also shown by Beltrami
to make perfectly good sense,
but this time on
a surface called a

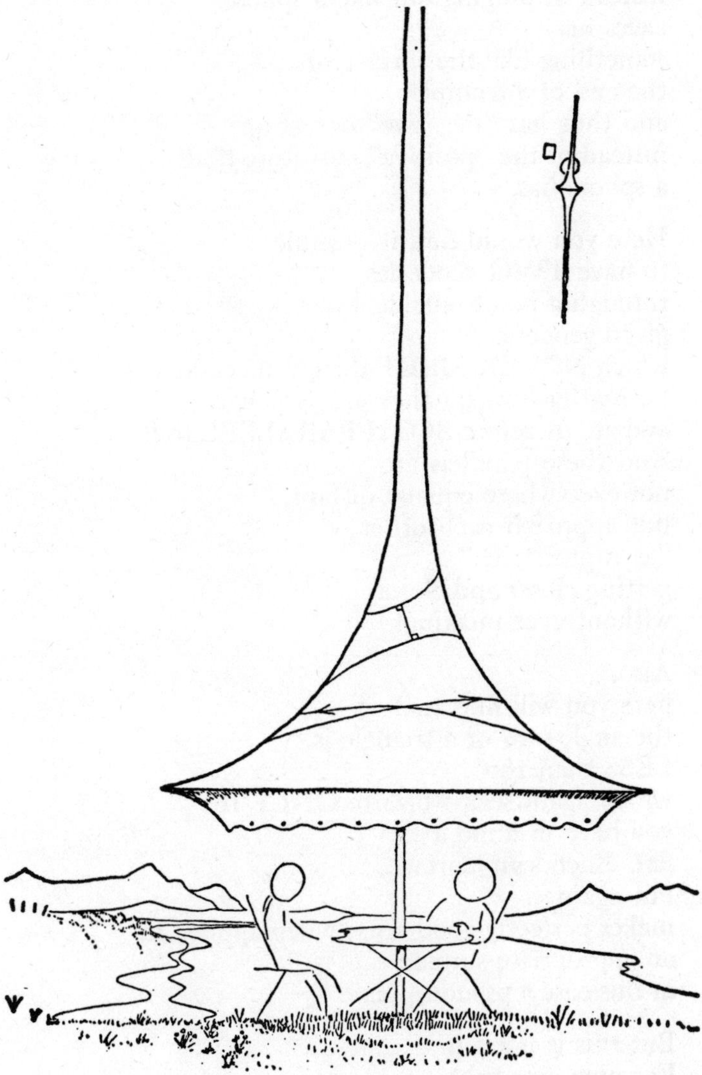

"PSEUDOSPHERE,"
which,
instead of bulging out like a sphere,
caves in—
something like the surface of
the end of a trumpet,
and thus has "negative" curvature
instead of the "positive" curvature that
a sphere has.

Here you would find it possible
to have TWO geodesics
through a point outside a
given geodesic
which NEVER MEET the given geodesic
no matter how far they are prolonged;
and are therefore BOTH PARALLEL to it.
And these parallels are
not everywhere equally distant,
but approach each other
"asymptotically,"
getting closer and closer
without ever meeting!

Also,
here you will find that
the angle-sum of a triangle is
LESS than 180°,
which again seems bizarre ONLY IF
you have in mind a
flat, Euclidean, surface,
but again
makes perfectly good sense when applied to
an appropriate surface,
in this case a pseudosphere.

But this is not all!
For now no doubt
your mind is limber enough to

accept geometries on still other surfaces,
like a surface with a hole in it,
called a torus ("doughnut" shaped),
and of course others with
two or more holes ("pretzel"),
etc., etc.

And you will now
quite readily accept
"FINITE" geometries,*
which have only
a finite number of points
(like 25 or 9, etc.)
instead of an infinite number,
as in Euclidean geometry
as well as in
both Non-Euclidean geometries
discussed above.

Thus you are now ready to accept
the really modern study of
geometry,
known as Topology—
which includes various geometries
for various important purposes.

And,
in addition to all this,
when Einstein,
in his famous Theory of Relativity
(discussed in a later chapter in this book),
studied the structure of
the Universe,
the actual world in which we live,
he found that
the most appropriate geometry

* See the paper "A Finite Geometry" by A. Church, published by The Galois Institute of Mathematics and Art, 624 A Third Street, Brooklyn 15, N.Y.

for this purpose
turned out to be
NON-EUCLIDEAN,
and not at all the
Euclidean geometry which
we all study in high school.
But of course this is NOT to say that
this old Euclidean geometry is
no longer any good—
on the contrary,
you now know that
it is as good as ever
(in fact better now than ever,
since some errors in it
have been corrected)
provided it is applied appropriately,
as, for example, on a
"flat" Euclidean surface.

And indeed this
NEW ATTITUDE TOWARD POSTULATES
(page 29)
has been applied also to
ALGEBRA,
and resulted,
as you might guess,
in a great many
algebras and arithmetics,
and even logics (in the plural!),
WONDERFUL and USEFUL,
as you will see in
the subsequent chapters in this book.

And so
to sum up:
the CHANGE OF ATTITUDE
toward POSTULATES
(the Great Discovery of 1826!)
has resulted in furnishing us with a whole

"Tool-kit"
instead of just
one "hammer" with its
LIMITED usefulness.

But of course,
whenever one produces
a new postulate set,
one has the RESPONSIBILITY of
making sure that the postulates
DO NOT CONTRADICT EACH OTHER
(for you cannot even play a game in which
the rules contradict each other,
can you?!).

And, further,
one has the RESPONSIBILITY of
following through to the
"consequences" (theorems),
as did the mathematicians mentioned above,
and not just "give out" with
a would-be set of postulates
without such responsible examination—
and just "walk away from it"—
as do some "leaders" in high places
who issue irresponsible press conferences
in which they sometimes say things which
may sound innocent enough at the time,
but which would be seen to lead to
most dangerous consequences
if properly and
RESPONSIBLY followed through!

Thus mathematics can tell us much about
PROPER HUMAN BEHAVIOR,
and is not to be used only as
a calculating tool,
like a machine!

II. "ETERNAL VERITIES"

1. Is Twice Two Always Equal to Four?

And so we have seen that,
in MATHEMATICS,
there are
NO "self-evident truths."
And, we shall soon see that
there are also
NO "eternal verities" in
MATHEMATICS!
To those people who,
though they may dislike mathematics,
still have a
"give-the-devil-his-due" attitude,
believing that
at least mathematics is
RELIABLE
since in it we have
such "eternal verities" as
TWICE TWO IS ALWAYS FOUR,
I say
you will soon see that
twice two is NOT always four,
that there are NO "eternal verities" in
mathematics—
AND YET—
do not get too discouraged,
for,
even though mathematics is
only a MAN-MADE enterprise,
still
man has done very well for himself
in this domain,

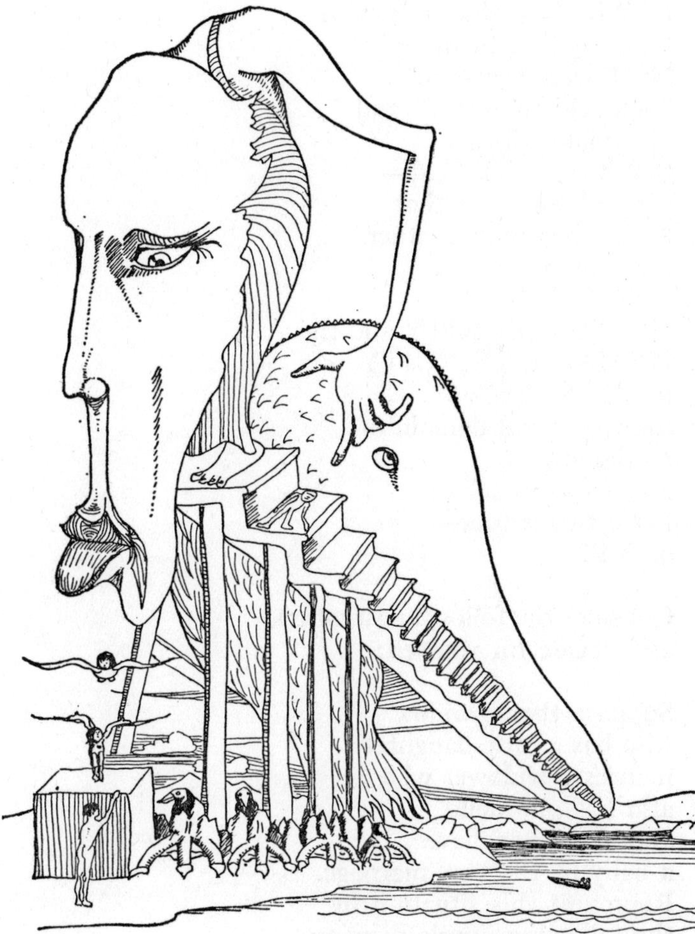

where he has
FREEDOM WITH RESPONSIBILITY—
and where
though he has learned the
HUMILITY that goes with
knowing that he does
NOT have access to
"Self-evident truths" and
"Eternal verities,"
that he is NOT God—
yet he knows also that
he is not a mouse either,
but a man,
with all the
HUMAN DIGNITY and the
HUMAN INGENUITY
needed to develop
the wonderful domain of
mathematics.
So let us go back to
twice two is four—
or is it?

Consider the following situation,
and decide for yourself:

Suppose that a widow
who has a little daughter
marries a widower who
also has one child,
and they then have
a baby by the new marriage.
Represent this situation by
the following simple diagram.
Let the two circles here represent
the "class" of HIS children and
the "class" of HER children—
then obviously the 1 child in

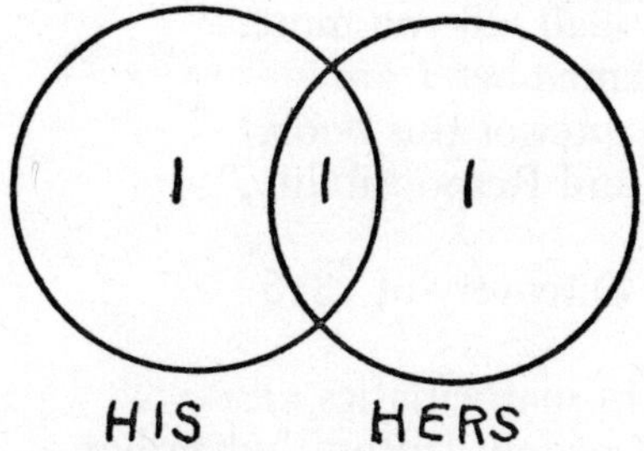

the overlapping portion represents
the new baby, and
the other 1's represent
HIS child and HER child BEFORE
this marriage.
Now,
it is quite easy to see that
although he has TWO children,
his own,
and she has TWO children,
her own,
yet TOGETHER
they have only
THREE children,
have they not?
That is,
here
TWO and TWO add up to
THREE,
NOT FOUR!
In other words,
you do NOT add here in the usual way,
but rather
in accordance with the meaning of
ADDITION of CLASSES
as defined in
BOOLEAN ALGEBRA.
This was the first of
the NEW Algebras,

of which I shall tell you more.
You will remember from
the first chapter of this book,
"Freedom and Responsibility,"
that
The Great Discovery of 1826
was that
Postulates in mathematics are
NOT "self-evident truths" but rather
MAN-MADE ASSUMPTIONS,
and therefore of course
subject to change by MAN!
And from this followed
various new
NON-Euclidean Geometries
which have served so many
useful purposes.
Remember?

Well, in THIS chapter
I wish to show you how
this same Great Discovery
influenced ALGEBRA also,
so that since then
there have grown up also
MANY NEW ALGEBRAS,
of which the first was
the Algebra of Classes,
known as Boolean Algebra,
mentioned above,
and developed by
George Boole
about 1850.
I shall not give you here
a complete set of the
Postulates for Boolean Algebra,
but will point out only
two of its important definitions,

"SUM" and "PRODUCT,"
which have a bearing on
the subject of
"Eternal Verities."
The SUM of two CLASSES, a and b,
is indicated in the following diagram

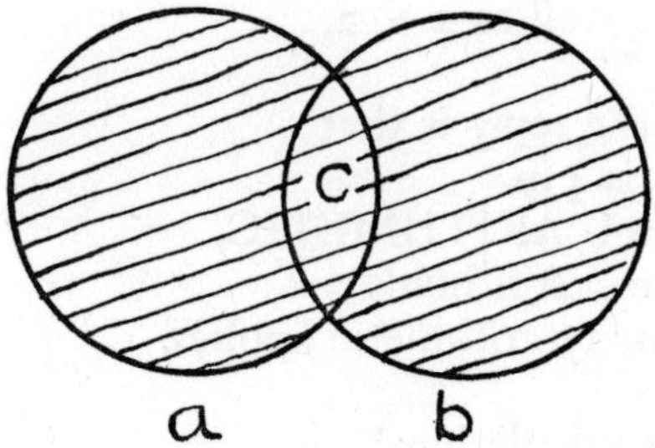

by the entire shaded portion—
that is why we got

$$2 \oplus 2 = 3$$

(on page 49)
when we were adding
the two classes of children,
HIS and HERS.
Note the use of $\oplus$ instead of
the usual $+$ of ordinary arithmetic,
since of course these two symbols
refer to DIFFERENT kinds of ADDITION.

Incidentally, may I say that
the "PRODUCT" of two CLASSES
is defined to be
their common part—
the part marked c in the above diagram.
So that,
in the situation on pages 48 and 49,
we have
$2 \otimes 2 = 1$ (not 4)!

And note of course $\otimes$ instead of
ordinary multiplication, $\times$.

Perhaps you will say
"Very clever, but
not to be taken too seriously,
for, after all,
$2 + 2$, as well as 2×2,
is still 4."
To which the reply is that
the answer is 4 in
ORDINARY ARITHMETIC,
and of course still holds
where ordinary arithmetic applies,
as in your
income tax calculations!
BUT you will admit that
there are many situations in
LIFE,
as in the case of the family on pages 48 and 49,
where ordinary arithmetic is
NOT the proper tool for getting
the correct answers!
And you will be surprised to know
how very many problems there are
which can NOT be solved by
ordinary arithmetic and algebra,
but rather by
Boolean Algebra,
or some other of
the many NEW algebras.
Take for instance
the tremendous advantage
that has been gained in
the study of Logic itself,
in which large volumes of
traditional logic
have been reduced to

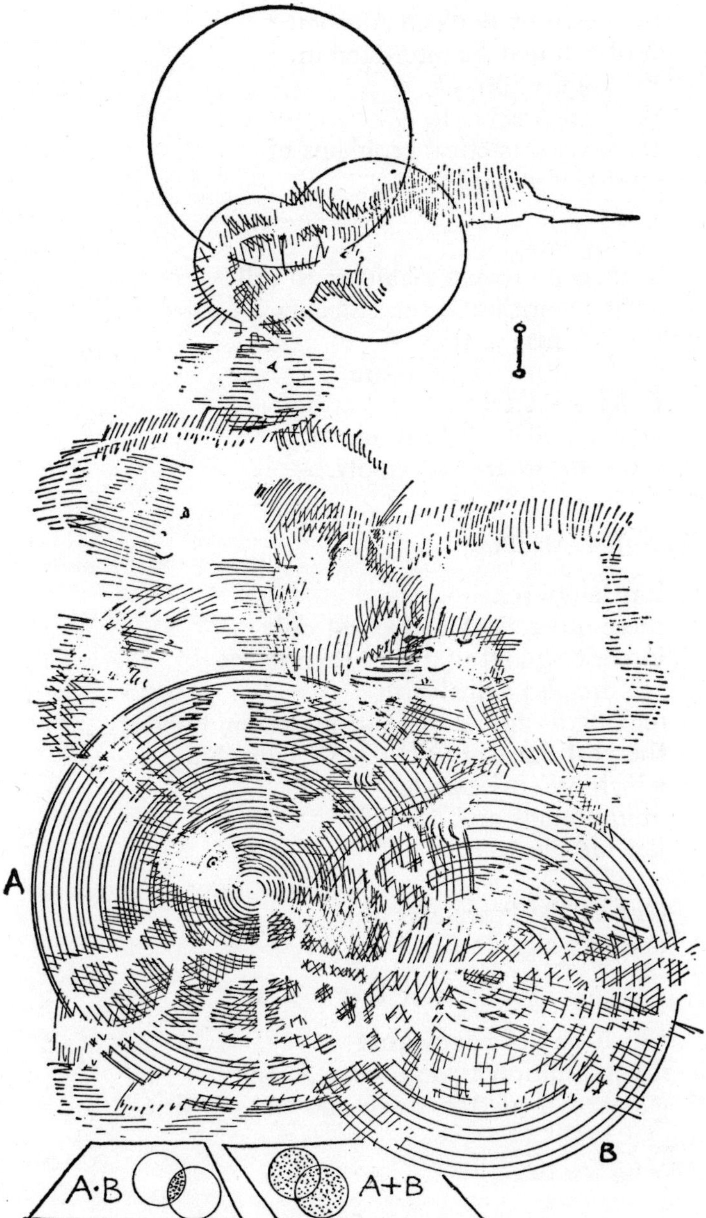

A
B
A·B
A+B

FIVE LINES
by means of Boolean Algebra! *
And you may be interested in
the applications of
Boolean Algebra to
some very practical problems of
electrical wiring! †

So you see
there is no reason whatever to call
ordinary arithmetic and algebra
the "eternal verities"—
for the other algebras are
JUST AS GOOD,
and indeed BETTER in
problems where they apply.

2. Other Algebras.

Furthermore,
these other algebras are not only
interesting and useful in themselves,
but are also valuable in
making us appreciate and understand better
the postulates of ORDINARY algebra
which otherwise seem
unimportant and "boring."
Let me illustrate:

One of the Postulates of
ordinary algebra is:

$$a \times b = b \times a$$

which says of course that
if you multiply two numbers, a and b,
it does not matter which one you put first,

* See Chapter III.
† See Chapter III.

the answer is the same in either case:
thus,

$$7 \times 11 = 77$$
$$11 \times 7 = 77$$
etc.

This is known as the
"Commutative Law of Multiplication."
Now I am sure
you are not thrilled by this—
it is so "obvious,"
it is just a bore to mention it!
But now prepare for a
SHOCK!

There is an algebra given by
E. V. Huntington in his
"Fundamental Propositions of Algebra" *
in which he defines $\otimes$ as follows:

$$a \otimes b = b$$

that is,
the product of two numbers here gives
the second one of the two
as the answer!
Thus here we have

$$4 \otimes 7 = 7,$$
$$5 \otimes 2 = 2,$$
etc.

so that you do not have to
learn any multiplication table here!
Youngsters would go for this in a big way,
wouldn't they?

* Published by
The Galois Institute of
Mathematics and Art,
624 A Third Street, Brooklyn 15, N.Y.

But of course Huntington
did not invent this algebra just
to please the children,
though he would not mind
their enjoying it!
He had a serious purpose in
inventing it—
look it up in his little book
mentioned on page 55.
Here I shall only point out
the following:
In this algebra, as you see,
we have

$$a \otimes b = b$$

BUT

$$b \otimes a = a$$

That is,
when you "multiply" two numbers here
you get two DIFFERENT answers
(b or a, as above)
depending on
which number you put first!

$$a \otimes b \text{ does NOT equal } b \otimes a,\text{—}$$

So that
that is
the Commutative Law for Multiplication
does NOT hold in this algebra
(contrast page 55)!
The knowledge that it is
POSSIBLE
to have algebras in which
this simple law does NOT hold,
emphasizes the fact that
ALL algebras are MAN-MADE,
with MAN-MADE postulates,

and that
in order to be able to use
ANY algebra,
you must of course know
what ARE its postulates,
and you must therefore
know and
STATE EXPLICITLY
whether, for example,
the Commutative Law

$$a \otimes b = b \otimes a$$

holds or not.
And so,
if it holds in ordinary algebra,
which it DOES,
one must of course SAY SO,
for it is far from "obvious" that
it MUST always hold,
as you saw above.
Thus,
by coming in contact with
VARIOUS algebras,
we learn to appreciate how important it is
to state explicitly
the Set of Postulates for
any particular algebra,
whether it be ordinary algebra
or any other—
for it is the Postulate Set which
tells us
what the game is
that we are trying to play.
And the fact that
the postulates are
FEW and SIMPLE,
far from being a bore,
is a WELCOME fact,

for then we realize that
the game is really a simple one—
and even ordinary algebra
thus loses its bad reputation of
being hard!

And you will be
amazed and delighted to see
the various algebras given by
E. V. Huntington
and to realize
the important connection they have
with ordinary algebra—
this will make the latter
much more interesting and alive
than the usual boring approach to
this beautiful subject.

Let me now make
very brief reference
to some other algebras
which are among
the most useful modern tools of thought:

(1) Take the system known as a
GROUP.
It is a very simple system,
for it has
only ONE operation,
"Multiplication" and
only FOUR postulates,*
and yet
by means of GROUP THEORY
many problems have been solved which
resisted solution for centuries—

* See *Galois and the Theory of Groups,*
by Lillian R. Lieber,
with drawings by Hugh G. Lieber,
published by
The Galois Institute of Mathematics and Art.

longer than a three-inch one which
is only a part of it—
this is "common sense"!

But you will soon see that
ven this statement about
e WHOLE and the PART
NOT an eternal verity!

ppose you know that
ertain stadium
a capacity of 50,000;
suppose that
a certain day,
ou look around in the stadium,
find that
seat is occupied by one person
there is no one standing,
vill surely agree that
are just as many people
e stadium
re are seats—
er words,
ow that there are
people present,
ough you have not counted them.
to say,
ts
case
of seats and
f people)
e-to-one correspondence—
person has a definite seat,
seat is filled by one person—
umber of individuals in each set
ME—
mber of people

for instance:
the famous old trisection problem *
is EASILY shown to be
IMPOSSIBLE of solution—
page the young and old would-be
angle-trisectors,
and those journalists who write
spectacular articles claiming
that some young "genius" in their town
has succeeded in performing
this impossible feat!
Such a "genius" would spend his time
to much better advantage
if he studied the Theory of Groups—
which would not only explain to him
why the trisection problem is impossible,
but would familiarize him with
one of the most important tools of
Modern Mathematics.
As E. T. Bell says: †
"Wherever groups disclosed themselves,
or could be introduced,
simplicity and harmony
crystallized out of comparative chaos.
The idea of a group
was one of the outstanding additions
to the apparatus of scientific thought
of the last century."

* The problem consists in trisecting the general angle, using only a straight edge and a pair of compasses. You must not select certain special angles, as, for example, a right angle, for this one CAN be trisected, but it is impossible to do this in GENERAL.

† See *The Queen of the Sciences*, by E. T. Bell, a great American mathematician.

is the SAME as
the number of seats.

Now let us apply this idea to
an INFINITE SET,
like the set of whole numbers:

1, 2, 3, ...

And now let us suppose that
these are seat numbers—
seat no. 1, seat no. 2, etc., etc.
Further,
let us assign these seats to
people who,
instead of being identified by their names,
are identified instead by NUMBER,
as follows:

Mr. 2, Mr. 4, Madame 6, Miss 8, etc.
(using EVEN numbers only here).

And, finally,
let us assign these seats to these people
in the following specific manner:

Seats	1	2	3	. . .	10	. . .	29	. . .
People	2	4	6	. . .	20	. . .	58	. . .

Note that in each case
the person's number is
double the seat number.
Thus it is easy to find out
that Seat 50 would be occupied by
the Person whose number is 100,
and,
Person 82 would be
in Seat 41,
etc.

Thus there is a
one-to-one correspondence between
the two sets,
the set of people and the set of seats,
and, therefore,
as you agreed on pages 63–64,
the number of people
is the SAME as
the number of seats.

All this sounds
simple and innocent enough,
BUT
it leads to
a most amazing conclusion!
For,
according to what has just been said,
the total number of numbers in
each row on page 64
is the SAME—
and yet,
the set of numbers in the second row,
being only the EVEN numbers,
is PART of
the set of numbers in the first row,
since the first row also contains
all the EVEN numbers, and lots more!
Thus we are obliged to admit that
the ENTIRE set of numbers

$$1, 2, 3, \ldots\ldots$$

is EQUAL to
a PART of itself,
namely,
the EVEN numbers only!
What then becomes of
the "eternal verity" on page 61 that

THE WHOLE IS GREATER THAN
ANY OF ITS PARTS?

Obviously then
it is NOT an "eternal verity" at all!
It is merely a Postulate which
holds for FINITE sets, but
NOT for INFINITE sets.

3. Plenty of Freedom—But NOT to LIE!

But you are probably
no longer shocked by this
since you know by now that
the POSTULATES which govern
any branch of mathematics
(like the RULES which govern any game)
depend upon the game,
and you may invent
any system or game
you wish.
And all this has
nothing whatever to do with
"'eternal verities,"
does it?

The enormous importance that
the Theory of Transfinites has
in ALL of mathematics
reads like a fairy tale.*

Let me inject
a WORD of CAUTION here—

* See "Infinity" by
Lillian R. Lieber,
with drawings by
Hugh G. Lieber,
published by Holt, Rinehart and Winston.

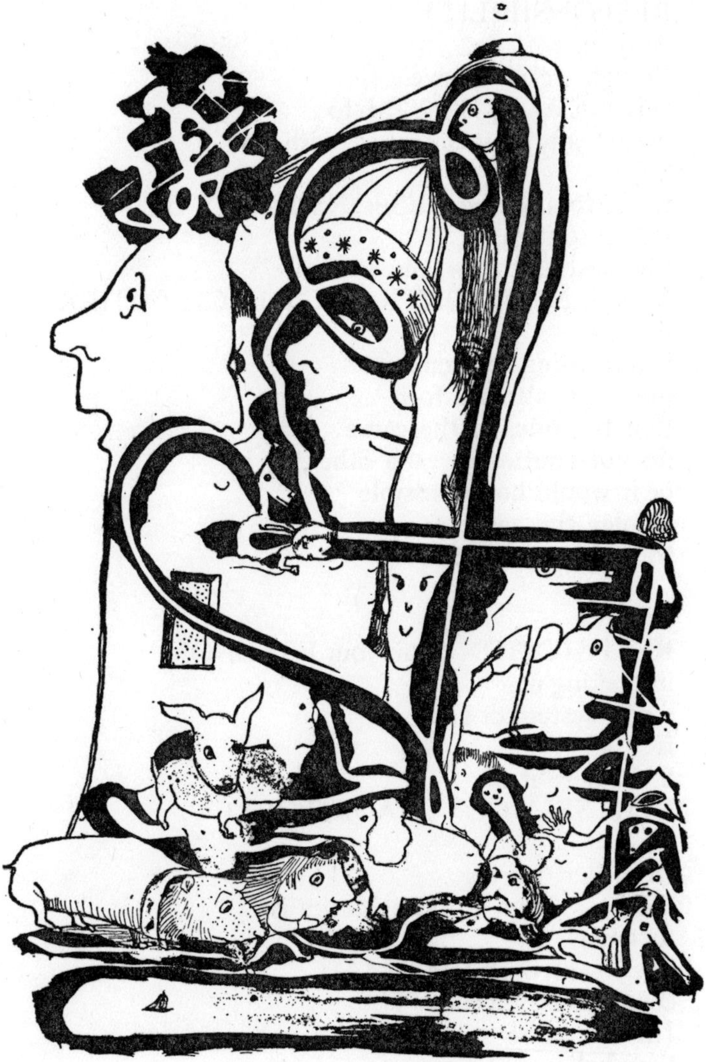

LEST WE FORGET that
FREEDOM in mathematics
MUST be accompanied by
RESPONSIBILITY:
And so,
though we now have
this NEW FREEDOM to
invent any postulates we wish,
STILL
we have also the RESPONSIBILITY of
seeing to it that
the postulates for any system
MUST NOT CONTRADICT EACH OTHER
just as,
in inventing any game
one must also see to it
that the rules of the game
do not contradict each other,
or it would be impossible
to play the game or
to use the system!

Thus, a
RESTRICTION upon your FREEDOM
in making up
a new system or game
is
to be sure it is
NOT SELF-CONTRADICTORY!

In other words,
the basic SIN in mathematics is
to LIE!
For if one postulate
contradicts another
(in the same system, of course)
they belie each other,
do they not?

Of course if you make up
a Postulate Set which
you wish to APPLY to the outside world,
as in the physical sciences,
or in the social sciences,
there is a FURTHER RESTRICTION in that
you cannot select the postulates quite
so arbitrarily,
for, they must
NOT ONLY be NON-SELF-CONTRADICTORY
but MUST ALSO be
APPROPRIATE to the FACTS involved:
thus, if it is
a Postulate Set for Physics,
it must be in agreement with
the experimental data—
remember the great triumph of
The Einstein Theory of Relativity *
which agreed with
so many experimental facts which
had not been previously explained!

And if you try to build
a Postulate Set for the
Social Sciences,
I hope you will agree with me that
one of the postulates should be:

LIFE FOR THE HUMAN RACE—

and then of course
the other postulates
MUST NOT CONTRADICT this
and lead to WAR!
But more of this in a later chapter.

* See Chapter IV.

And so I submit that
mathematics can shed light on
ideas like
FREEDOM with RESPONSIBILITY
as against LICENSE!
That mathematics can clear up the error of
mistaking man-made rules for
"eternal verities";
that it can show the
complete inadequacy of
the BIG LIE as
a basis for
any system of thought!
And therefore I submit that
mathematics is full of
HUMAN VALUES,
and not a machine for
counting fast bucks!

It must be obvious even from
this short discussion
that
mathematics is full of
HIGH-POWERED IMAGINATION,
without resorting to the
BIG LIE!
And therefore it is quite
unnecessary and undesirable
to encourage lying as a method of
developing the "imagination,"
as is advocated by some people who
are familiar only with
the picayune and unconvincing
"imagination" of the
LIAR,
big or small,
and who do not know about
the truly great imagination it takes

to invent
NEW and DEEPLY SIGNIFICANT
mathematical ideas!
These are not only
interesting and useful,
but,
by their very HONESTY,
give us a sense of
SECURITY
which we so desperately seek,
and which
no liar can give us for long—
for when we find him (or her) out,
and this is sure to follow as
the night the day,
our disgust is deep
and our faith in human nature
is terribly shaken.

Perhaps this is
what is bothering our youth today.
Perhaps they feel that
they have been lied to
too often,
and do not have
sufficient opportunity
to see HUMAN NATURE as it functions in
mathematics—
with
IMAGINATION,
HONESTY,
SUCCESS,
SECURITY.
We must not offer them less!

III. LOGIC

1. A Way of Life

Perhaps you are impressed by the fact that
now we have
MANY geometries instead of only one
(the Euclidean Geometry we studied in school),
and MANY algebras (see Chapter II)
instead of only one,
and perhaps you are even willing to admit
that these are USEFUL—
but are you saying
"What is the good of it all when,
with all this mathematics
and logic
and science,
MAN is really
NOT improving his condition,
but is steadily going DOWN hill
toward ever more horrible wars
with ever more dreadful weapons,"
and indeed you may be thinking that
it is that very growth of
mathematics and
logic and
science
which is the CAUSE of this
deterioration in Man's condition?!
Is it not the Scientists themselves
who are leading us down this
dreadful path?
And therefore what we
desperately need is
NOT more mathematics, etc.,

but rather
a sense of VALUES,
a feeling for what is GOOD and what is BAD
for the human race,
and such a sense of VALUES
is not to be found in
the study of mathematics
but rather in
the study of "the humanities."

Now,
without in any way belittling
"the humanities,"
for I fully agree that
the great Poets, Musicians, Artists,
Religious Leaders,
do indeed give to Man
a sense of his
HUMAN DIGNITY and HUMAN VALUES—
but what I am maintaining is that
he can ALSO get these Values from
a PROPER study of mathematics,
and get it EVEN MORE CLEARLY here.
Indeed, if you look back at
what has already been said
in Chapters I and II,
you will find that
from a PROPER study of mathematics
we can see that

(1) whereas FREEDOM is GOOD
(since it permits Man to be
more CREATIVE and
thus learn to ADJUST to
an ever changing environment,
by learning new ways to
solve the various problems
which confront him),

yet he is NOT free to do
just ANYTHING—
for instance,
he cannot build his life on
the theory of
the BIG LIE
(as Hitler advocated!);

(2) furthermore
a PROPER study of mathematics
shows us the tremendous importance of
our ATTITUDES,
how we can change
not only our environment,
but OURSELVES,
by getting rid of
the most deep-seated
PREJUDICES,
which we used to consider to be
"eternal verities,"
but now know are
indeed ONLY PREJUDICES
and CAN be CHANGED and ADJUSTED
as occasion demands this
for our own GOOD;

(3) that a proper study of mathematics
shows us also that
changes are not made
lightly and flippantly,
but only after thorough examination
and deep thought
(NOT BY VIOLENCE!),
thus showing us
how to be a
CONSERVATIVE PROGRESSIVE;

(4) and thus what looked like
FAILURE for a long time

eventually turned into a
wonderful SUCCESS *—
if we give it the necessary
TIME and THOUGHT and PATIENCE
and do not resort to
either VIOLENCE
or to the other extreme of
just STANDING STILL
without making any progress at all!

These are all very valuable
lessons for living,
do you not think so?

But this is only the beginning:
before you have finished reading
this little book,
perhaps you will agree that
mathematics can really be
a way of life.

2. SAM

But you are still troubled by
the scientists who make
nuclear bombs
and who are working on
CBR (chemical, biological and radiological)
WARFARE,
are you not?

What "VALUES" have these scientists
derived from their study of
mathematics and
logic and
science?
I believe that the difficulty here is

* Remember "The Great Discovery," Ch. I.

one of SEMANTICS—
what is the meaning of the word
"Scientist"?
This is really a very loose term,
and means anything from
a maker of bombs to
the great philosophic men
(like Einstein, Newton, and others)
who abhor the very idea of
making bombs,
and feel that
a PROPER study of Science
is basically concerned with
LIFE for the human race
(see Einstein's statement in Note 1, p. 142)
and is basically
OPPOSED TO DESTRUCTION:

Do you remember Einstein's famous
TELEGRAM TO THE PEOPLE
in which he WARNS of *
the overwhelming responsibility
to harness atoms
for BENEFIT of MANKIND
and NOT for humanity's destruction!
(see Note 1, p. 142)

We shall come back to this
important question
in Chapter V.

In the meantime,
to avoid the confusion arising from
the various meanings of the word
"Scientist,"
let us avoid this word altogether,
and make up a new word

* See Note 1, p. 142.

which we can define
without the confusion mentioned above:

this word is SAM,
originally derived from
Science, Art, Mathematics,
but better thought of as follows:

the S represents our "observations"
(what we see and hear with our senses
and with the aid of instruments),
the A represents our "intuition"
(which helps us to select
the basic postulates for a system
to deal with these observations—
of which we shall say more later),
and the M represents
the deductive processes by which
valid conclusions are drawn
from the postulates.

Thus the total word SAM
is DEFINED to be
one who uses
the BEST PROCEDURES of
mathematics, logic, and science
for the BASIC demand of
LIFE FOR THE HUMAN RACE.

And so "SAM" better represents
the idea of a "true" Scientist,
meaning one who is
UNALTERABLY OPPOSED TO DESTRUCTION,
as being basically CONTRADICTORY
to the SPIRIT of Science.

But, as I said before,
more of this in

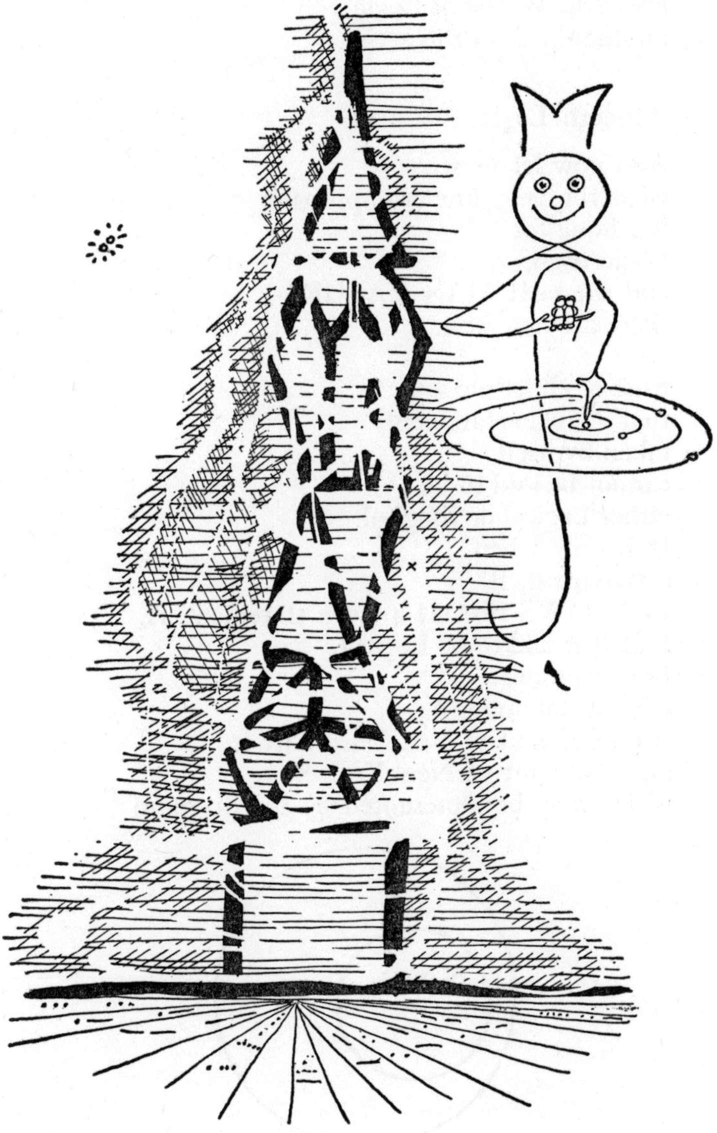

Chapter V.
And for more about SAM,
see *Mits, Wits and Logic*
another book in this series.

3. Modern Logic

And now let us see
what the new, modern approach
has done for
Logic itself,
and what HUMAN VALUES
there are here also.

First of all let me say
that a single statement,
taken by itself,
cannot be said to be
either Logical or Illogical.
It is only when
a statement, B,
FOLLOWS FROM a given statement, A,
that B is said to be Logical.
For example,
take as statement A:
An inhabitant of New York City is
an inhabitant of New York State,
which may be represented by the diagram:

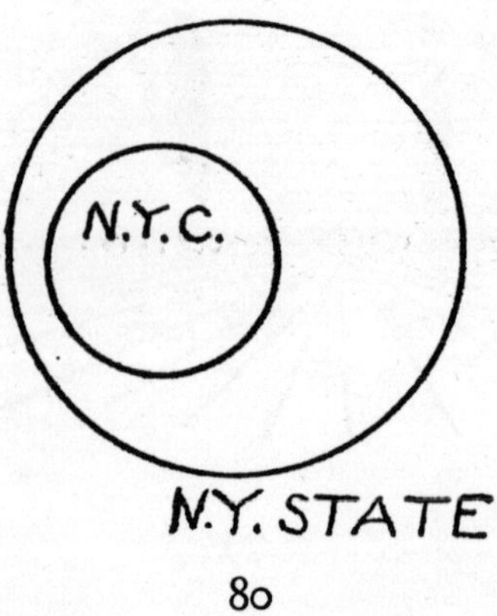

Then B FOLLOWS from this, namely:
If a person lives OUTSIDE of
New York State then
he lives also OUTSIDE of
New York City.
(Note that this may be seen from
the same diagram, above.)
Then B is Logical here.

Although diagrams are sometimes
convenient for this purpose,
they would not serve as a
PROOF!
a proper PROOF of a proposition
is to show,
step by step,
how it is derived from
the basic set of postulates of
a given system.
We shall soon illustrate this.

As was stated in Chapter II,
it is the new Boolean Algebra
(the Algebra of Classes)
which applies to Logic,
and we shall therefore give you
in this Chapter
a complete list of postulates
for this algebra,
and illustrate how
propositions of Logic are "proved"
from the postulates.

But first let me repeat what
was said in Chapter II:
one of the remarkable achievements
of Boolean Algebra is
that it has simplified
traditional logic

to such an extent that
the venerable subject of logic,
first invented way back,
centuries ago,
by Aristotle himself,
and carried on through the ages
by many learned scholars,
can now be summed up in
ONLY FIVE LINES!
If you are interested in
the intricate details of
the old style logic
(which, by the way,
is still being taught
in many college courses in logic!)
with its many kinds of
SYLLOGISMS,
"categorical"
"hypothetical,"
"disjunctive,"
and the ridiculous "mnemonic" of
"artificial words,"
which Jevons,
a well-known British logician
in 1881 (!)
described as a
"device of a barbarous and
wholly unscientific character"—
and yet is still being taught to
students of logic
NOW in 1961!—
I say, if you are interested in
these gruesome details,
you can look it up
in the book referred to on p. 80
of this Chapter.

But here let us get down to business
and do it in the MODERN WAY,

by starting with
a set of POSTULATES for
Boolean Algebra
(the Algebra of Classes).

But first
a few brief explanations
are in order:

(1) the "elements" in this algebra are
"CLASSES"
(NOT numbers, as in ordinary algebra);

(2) the SUM and PRODUCT
of two classes
are defined on p. 51 of Chapter II.
We repeat them here
for your convenience:

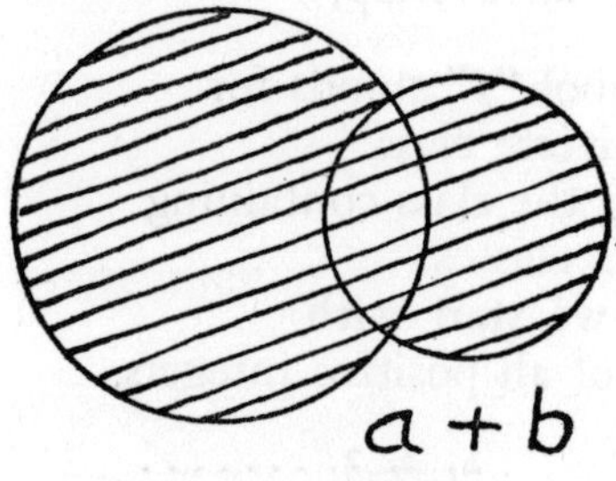

the SUM of classes a and b
is the complete shaded area,
as shown;

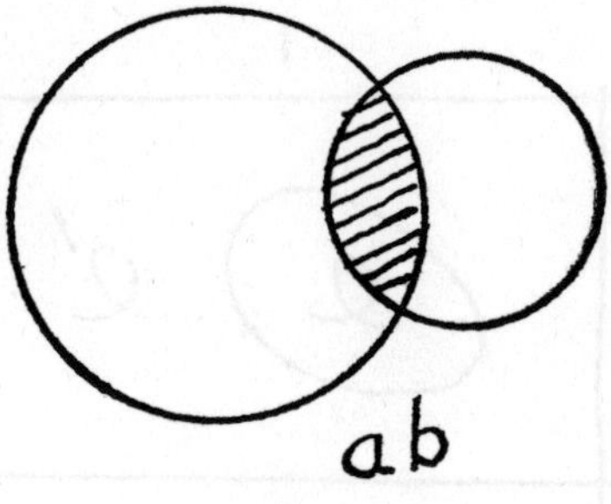

the PRODUCT of classes a and b
is the area COMMON to both, c.

(3) the symbol "o" stands for
the "null" class—
for example

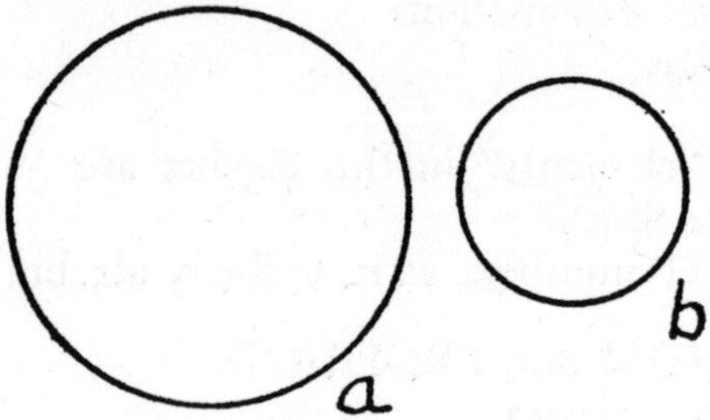

here ab = o
since the two classes, a and b,
have nothing in common;
they do not overlap;

(4) the symbol "1" stands for
the "universe class,"
meaning the class containing
ALL the classes:
thus if we start with
the set of all positive integers,

1, 2, 3, ...

and represent the entire set
by the rectangle in
the following figure:

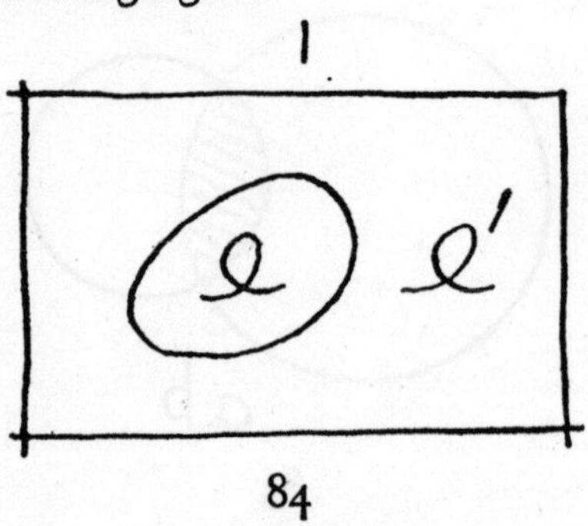

then if the class e contains
all the EVEN numbers only,
then obviously
all the ODD numbers are in
class e′,
which is called
the "complement" of class e,
and of course

$$e + e' = 1;$$

(5) some sets of elements have
"closure" under a certain operation—
thus, for example, to say that
the set of EVEN numbers
has closure under addition
means that
the SUM of any two EVEN numbers
is itself an EVEN number;
whereas to say that
the set of ODD numbers
does NOT have closure under addition
means that
the SUM of any two ODD numbers
is NOT itself an ODD number—
thus $3 + 7 = 10$ (an EVEN number).

4. Postulate Set for Boolean Algebra.

Let us now give a list of
POSTULATES (basic rules)
for Boolean Algebra:

from the definitions of
SUM and PRODUCT of two classes (p. 51)
we may say that
the Algebra of Classes has
"closure" under addition and also

"closure" under multiplication,
which may be expressed by

(1) $a + b = c$, a unique class,
i.e., adding two classes of the system
gives a result which is
itself a class in this
universe class.

Similarly

(2) $ab = c$, another unique class,
says that
the PRODUCT of two classes
is itself a class in this universe class
(even if the PRODUCT happens to be
the "null" class, as on p. 84).

Further

(3) $a + 0 = a$
(i.e., adding the null class to
a given class, a,
leaves a unchanged)

(4) $a \times 1 = a$
(i.e., the part common to
any class, a, and
the universe class, 1,
is the class a itself, thus:

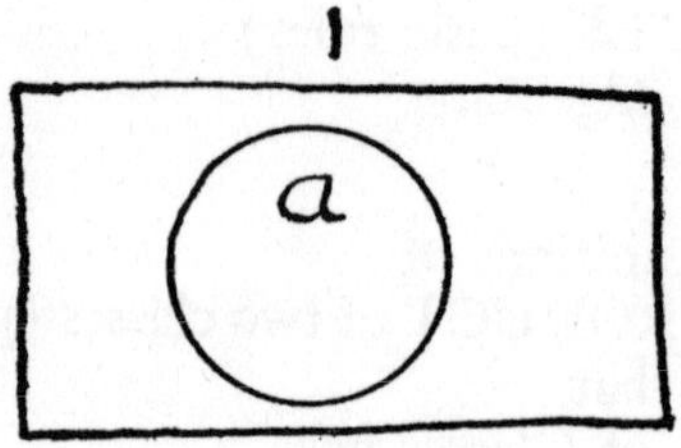

(5) $a + b = b + a$;

(6) $ab = ba$;

(7) $a(b + c) = ab + ac$;

So far, these seven postulates
hold also in
the ordinary algebra of
real numbers:
i.e.,
the two "closure" laws,
for ADDITION and MULTIPLICATION,
the number o which,
when ADDED to any real number
leaves that number unchanged,
and the ordinary "unity" element,
the number 1, which
when MULTIPLIED by any real number,
leaves that number unchanged;
then (5) and (6) are
the usual "commutative laws"
for ADDITION and MULTIPLICATION,
so familiar to us in
the algebra of real numbers;
and of course (7) is
usually referred to as
the "distributive law"—
but had better be described
by saying that
MULTIPLICATION is
DISTRIBUTIVE over
ADDITION,
as shown in (7).

But now we find in
Boolean Algebra,
ANOTHER DISTRIBUTIVE LAW—
in which

ADDITION is
DISTRIBUTIVE over
MULTIPLICATION, thus:

(8) $a + (bc) = (a + b)(a + c)$
which is certainly NOT true for
real numbers,
as we see in the following illustration:

$5 + (2 \times 3) \neq (5 + 2)(5 + 3)$,
since the left-hand side
is equal to $5 + 6$ which equals 11,
whereas the right-hand side
is equal to 7×8 which equals 56,
and $11 \neq 56$,
does it?

But (8) DOES hold for CLASSES,
as shown in the following diagram:

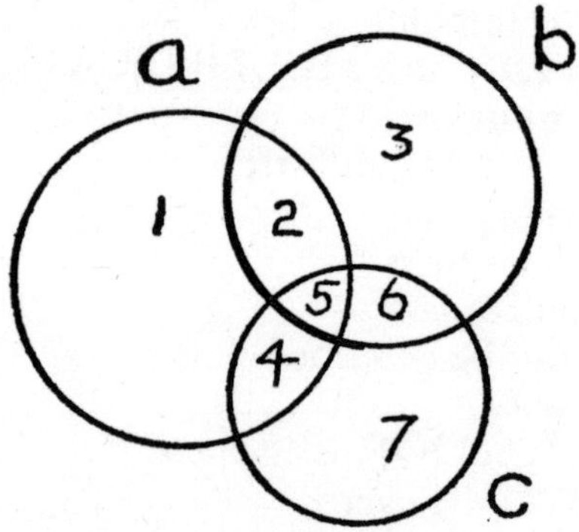

here a equals the areas 1245
and bc equals the areas 56;
therefore
$a + bc = 12456$,
AND
$a + b = 123456$ and

a + c = 124567,
so that
(a + b) (a + c) which is
the part COMMON to a + b and a + c
is obviously

12456,

the SAME result as was obtained
from a + bc.

So that here
the STRANGE DISTRIBUTIVE LAW (8)
really does hold.

Further,

(9) a + a′ = 1
(see the definition of
"complement" on p. 85);

(10) aa′ = 0
since there is nothing in common
to any class a and its complement a′:
as, for example,
the class of EVEN numbers
has nothing in common with
the class of ODD numbers;

(11) there are AT LEAST TWO
DIFFERENT
elements in the system.

These eleven postulates constitute
the Algebra of Classes,
known as Boolean Algebra,
which suffices to express
all of Traditional Logic in
ONLY FIVE LINES—

if you are interested in
seeing just how this is done,
you can find the necessary details
in Chapter XVII of the book
previously mentioned,
Mits, Wits and Logic.

5. An Amazing Practical Application

But here I would rather
show another illustration
of a marvellous
application of Boolean Algebra
to a very "practical" problem
made by F. E. Hohn
in the Bell Telephone Laboratories.*
Hohn showed how to use
Boolean Algebra
to effect tremendous ECONOMIES
in electrical circuits!

to understand this
a few definitions will first be necessary:

Let a, b, c, etc., represent switches;
let 1 represent
the condition of a switch being
"closed,"
and 0 the condition of a switch being
"open."

Further,
let the SUM of two switches mean
connecting them "in parallel,"
as in Figure 1:

* See "Some Mathematical Aspects of Switching"
by F. E. Hohn,
in the *American Mathematical Monthly,*
vol. 62 (1955) pp. 75–90.

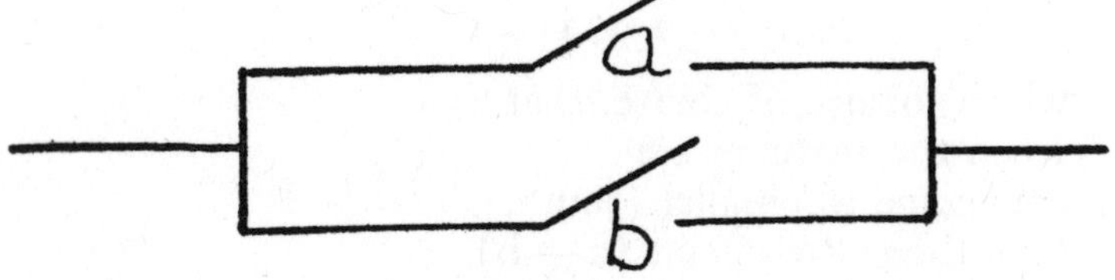

a + b

Fig. 1

and let the PRODUCT ab mean
connecting them "in series,"
as in Fig. 2:

ab

Fig. 2

Now, since any switch can be only
"open" or "closed,"
we can have only
the following possible combinations
for a pair of switches:

SUMS

$0 + 0 = 0$

$1 + 0 = 1$

$0 + 1 = 1$

$1 + 1 = 1$

PRODUCTS

$0 \times 0 = 0$

$1 \times 0 = 0$

$$0 \times 1 = 0$$
$$1 \times 1 = 1$$

which means, of course, that
when the switches are
connected in parallel (sum),
then the entire circuit $(a + b)$
is "open" $(= 0)$ only if
BOTH switches are "open" $(= 0)$!

And when the switches are
connected in series (product),
then the circuit is "open" $(= 0)$,
that is, the current CANNOT flow
in ALL cases, EXCEPT
when BOTH switches are "closed" $(= 1)$.

Both tables given on pp. 91–92,
it should be noticed,
are in agreement with
the POSTULATES for Boolean Algebra (pp. 86–89).
(Some of them follow directly from
the postulates themselves,
and the others can be
easily proved from them.)

Finally,
a' is defined as shown in Fig. 3:

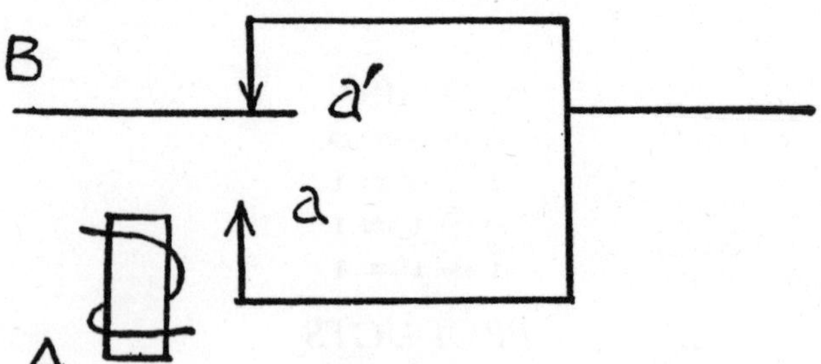

Fig. 3

by a judicious INTERPRETATION of
the basic terms (like 0, 1, sum, product, etc.),
and of course many, many other
APPLICATIONS
are possible—
you may even solve
some difficulty of your own,
entirely different from the one given above,
by making a proper, but different,
INTERPRETATION
of the basic terms—
IT IS UP TO YOU!

6. But Are There Any HUMAN VALUES here?

Well, it is obvious that
whenever we behave like
SAM,
using the data (S),
a Postulate Set
(similar to the one given above,
which demanded the intuition, A,
of the great George Boole),
and deducing (M) from the Postulates
various equivalent statements,
we make the kind of
PROGRESS
that is so essential
whenever we have to
THINK STRAIGHT,
and not just go around in circles
as so many people do who
are not acquainted with
the POSTULATIONAL METHOD
described above.
And of course
we must always bear in mind
that SAM is, BY DEFINITION,

one who uses
the BEST procedures of
mathematics, logic, and science
for the BASIC demand of
LIFE FOR THE HUMAN RACE,
and therefore CANNOT,
without CONTRADICTION,
be used for the
DESTRUCTION of HUMAN LIFE.

See again
Einstein's TELEGRAM TO THE PEOPLE,
part of which is given in Note 1 on p. 142.

7. Behind the Scenes in Science!

In other words,
the HUMAN VALUE of the
POSTULATIONAL METHOD
described above
will be evident
if we can set up
a proper POSTULATE SET
for HUMAN BEHAVIOR—
which I shall attempt to do
in Chapter V.

Furthermore,
I shall there
attempt to show, also,
how intimate is the relationship
between such a Postulate Set
and
SCIENCE and MATHEMATICS—
not only by
the very idea of using a
POSTULATE SET,
but also by

showing how
such a postulate set
is really
BEHIND THE SCENES
in Science (and mathematics) itself—
and may be called
METASCIENCE (and METAMATHEMATICS),
(see Note 2 p. 144.)
Without which
Science itself,
as well as Mathematics,
could not exist!

From this point of view
it is obvious that
anyone engaged in the making of
destructive weapons,
without a deep UNDERSTANDING
of the METASCIENCE,
should really not be called
a Scientist at all,
for he is just a mechanical ROBOT—
a mere GOLEM *—
who MUST be controlled by
SCIENCE in its DEEP SENSE,
as discussed above,
and will be discussed more in detail
in Chapter V.

* A Golem, in Jewish legend, is
an artificial man,
especially one created by
the cabalist Rabbi LOEW of Prague
at the end of the 16th century:
hence, an automaton
(cf. Frankenstein)!

IV. EQUAL RIGHT AND EQUAL SUCCESS

Thus far we have seen
that
many and various branches
of mathematics
are possible
and that they can all
co-exist,
each useful for
specific and various problems.

But how about
the actual PHYSICAL WORLD—
are we as free there
to make up
various postulates,
or are we held down there
by the actual physical "facts"
(in addition to the
restrictions imposed by the
METAMATHEMATICS
(See Note 2, p. 144.)

The answer is
"Yes, of course,
BUT you will soon see
how even here
there is
considerable freedom for
various observers
to study the universe
from their own viewpoints

with
equal right and equal success,
as Einstein expressed.

1. What is Democracy?

Many people speak of
democracy,
but,
as is true of
many other words,
the meanings are not always the same
and not always clear.

Thus, to some
"democracy" means that
everyone shall have
the right to decide
every issue;

to others it means that
everyone shall have
the right to
SELECT REPRESENTATIVES
who will then be the ones
to decide the issues;

and there may be still other
definitions of "democracy"—

for instance,
everyone should have "equal rights"—
but equal rights for WHAT?

Is it the right to
grow and develop
and do one's best
for humanity?

Or is it the right to be "wrong"?

Or is it rights of the
average man VERSUS the "egghead"?

Is it anti-democratic to
DIFFERENTIATE between individuals—
is it fair to "draft" everyone
into the infantry
instead of giving special people
special training in
science,
music,
art,
etc., etc.?

Now let us see whether
any light can be shed on this point
by an examination of
issues which arise in
Modern Mathematics and
Modern Physics.

2. Invariants under a Transformation.

This illustrates very well how
democracy functions in
mathematics and science.

Suppose, for example,
that you wish to measure
the distance between two points,
A and B,
but there is an obstruction
(like a mountain or a lake)
between them,
so that you cannot go directly

from one of the points to the other,
as shown in Figure 1:

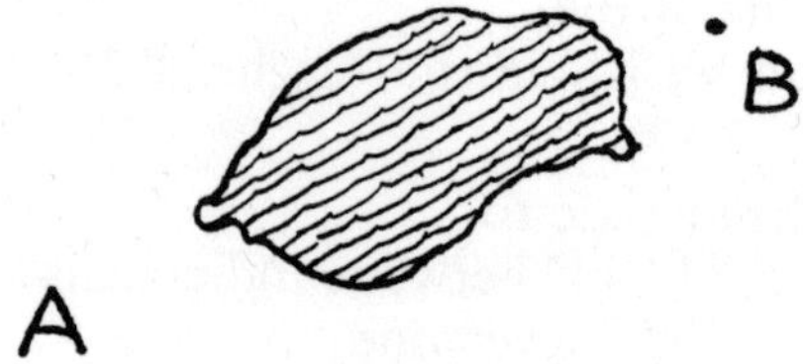

Fig. 1.

A clever man,
call him Mr. K
(This symbol was used by Einstein
in his
Special Theory of Relativity,
way back in 1905!),
would suggest that you can
solve the problem
in the following way,
using Figure 2:

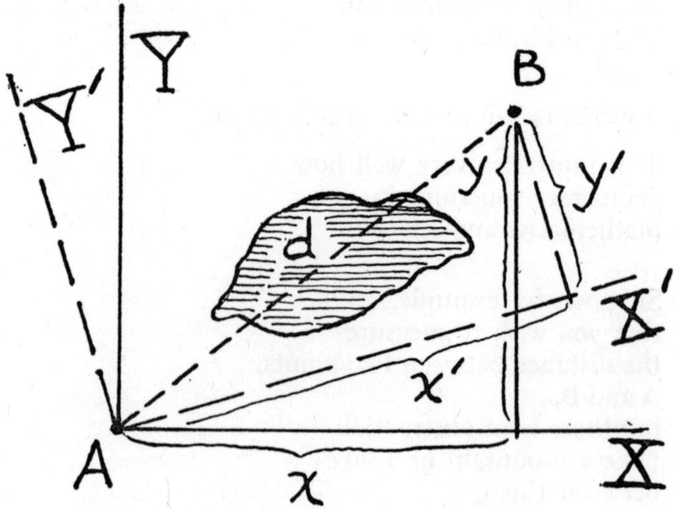

Fig. 2.

Draw the line AX
(avoiding the obstruction)
and AY perpendicular to AX
(these are called the axes
or a "coordinate system")
and then through B
draw another line, y,
parallel to AY as shown;
and now representing
the distance between A and B by d,
we have,
by the well-known
Pythagorean Theorem,

$$d^2 = x^2 + y^2$$

and therefore

$$d = \sqrt{x^2 + y^2} \quad (1)$$

Thus,
by measuring the distances x and y,
you can then CALCULATE
the REQUIRED distance d
by means of formula (1),
without measuring it DIRECTLY.

Of course an equally clever man,
say, Mr. K′,
could have drawn the line AX′
(instead of AX),
and AY′ perpendicular to AX′
(thus using a different coordinate system,)
and then drawn a line parallel to AY′,
namely y′—
and he could EQUALLY well
find the distance d
by the Pythagorean formula

$$d = \sqrt{(x')^2 + (y')^2}$$

Note that even though

K and K′ have DIFFERENT
coordinate systems,
that is,
$x \neq x'$ and
$y \neq y'$,
still they do agree on SOMETHING—
in this case, the value of d;
and we say that

$$\sqrt{x^2 + y^2}$$

is an INVARIANT under the
rotation of axes shown in Fig. 2.

And even if still another
clever man, Mr. K″
used a coordinate system
AX″ and AY″,
which are not even perpendicular
to each other,
as shown in Fig. 3,
and then drew a line through B
parallel to AY″,
there is another formula
(NOT the Pythagorean Theorem,
but one known to anyone who
has studied a little Trigonometry),
namely,

$$d = \sqrt{(x'')^2 + (y'')^2 - 2\,x''y''\cos\theta}$$

where θ is the angle shown in Fig. 3,
by means of which
the distance d can be CALCULATED
by different but
appropriate measurements
here.

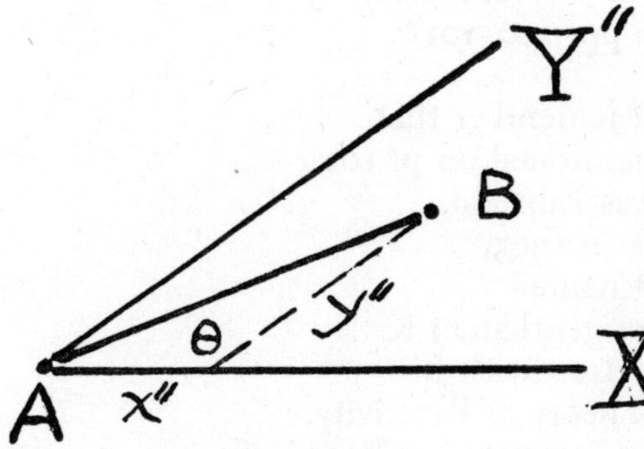

Fig. 3.

And of course there are
many, many other "coordinate systems"
which can also serve the same purpose,
all giving the SAME value of d.

But what has all this to do with
DEMOCRACY?!

Well, you can easily see that
whereas K, K′, K″ and
all others interested in
the problem of finding the
distance between A and B,
can do so with
"EQUAL RIGHT AND EQUAL SUCCESS,"
even though they all use
DIFFERENT COORDINATE SYSTEMS,
that is,
even though
each one approaches the problem
in his own way!

Now, is not this
the ESSENCE of DEMOCRACY?

And is it not CLEARER THAN
the assortment of

definitions of Democracy
given on pp. 100–101?

You will remember that
it was mentioned on p. 102
that it was Einstein,
way back in 1905,
who first named
the clever gentleman K
in connection with his
Special Theory of Relativity.
If you are interested in knowing
how this came up in
Einstein's Theory of Relativity,
you may wish to look it up in
another book in this series,
The Einstein Theory of Relativity
by Lillian R. Lieber
with drawings by
Hugh G. Lieber,
published by
Holt, Rinehart and Winston.

Let us now see whether

$$\sqrt{x^2 + y^2}$$

is still an INVARIANT
under a ROTATION of AXES in
THREE dimensions,
as shown in Fig. 4:
here the point B has
THREE coordinates,
namely, x (= AD), y (= DC), and z (= BC)
where z is perpendicular to
the PLANE of AX and AY,
and, as you probably know,
here $d^2 = x^2 + y^2 + z^2$
and therefore

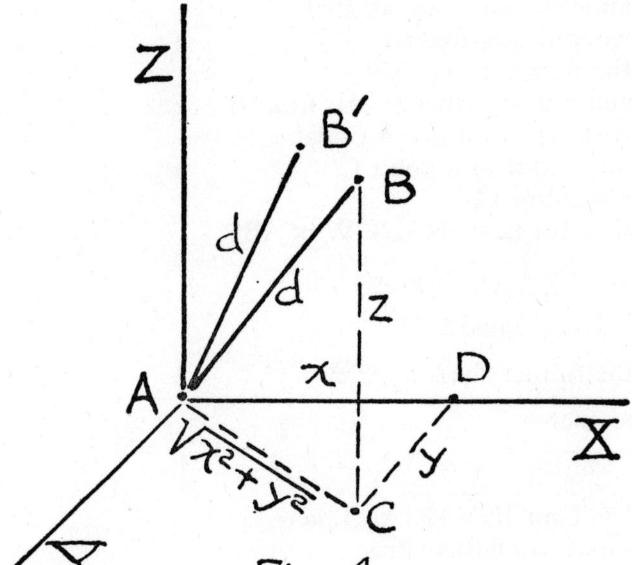

Fig. 4

$$d = \sqrt{x^2 + y^2 + z^2}.$$

Now, if instead of
revolving the axes
as we did before,
we revolve AB to a new position AB′,
the result is the same.
In other words,
the INVARIANT here is

$$\sqrt{x^2 + y^2 + z^2};$$

But how about $\sqrt{x^2 + y^2}$?

You may regard $\sqrt{x^2 + y^2}$ as
merely the SHADOW of d
on the XY plane,

and you can easily see that
when d is moved to
the new position AB′,
and a new perpendicular from B′
is dropped on the XY plane,
meeting it at a point C′
instead of C,
then the new SHADOW of AB′,
namely $\sqrt{(x')^2 + (y')^2}$,
is NOT equal to
the former shadow, $\sqrt{x^2 + y^2}$,
so that

$\sqrt{x^2 + y^2}$ is

NOT an INVARIANT here.
Thus, in mathematics,
we do not look for
ETERNAL VERITIES,
but make a much more modest request:

to find what are the INVARIANTS
under a SPECIFIC transformation.

And, indeed,
when Einstein re-examined
the physical universe
because it became necessary to do so
since new data produced by
Michelson and Morley (about 1887)
no longer fitted in to
the Newtonian system of the universe—
I say, when Einstein was thus
forced to re-examine
the physical universe,
he suggested first of all that
the ELEMENTS to be used

should NOT be POINTS
(as in Newtonian physics),
but rather EVENTS—
and of course an event is
a FOUR-dimensional element
since it occurs at a certain
longitude,
latitude,
altitude and
time,
or, more generally, it has
FOUR coordinates, namely,
x, y, z, τ
(τ being not the "time" itself
but a quantity related to it
which Einstein found to be useful),
and therefore proceeded to
look for the INVARIANTS
UNDER A SPECIFIC TRANSFORMATION,
known as the LORENTZ TRANSFORMATION *—
if you are interested in the details of this,
you can find them in the book
referred to on p. 106.
The Einstein Theory of Relativity
And he found that

$$d = \sqrt{x^2 + y^2 + z^2 + \tau^2},$$

and that

$$\sqrt{x^2 + y^2 + z^2 + \tau^2} = \sqrt{(x')^2 + (y')^2 + (z')^2 + (\tau')^2}$$

which means that
K, using the coordinates

$$x, y, z, \tau$$

and K′, using a different set of axes
and getting the coordinates

$$x', y', z', \tau'$$

* See Note 3, p. 147.

BOTH AGREE on the SAME value of d,
IN SPITE OF THEIR DIFFERENCES:
NAMELY

$$x \neq x', \ y \neq y', \ z \neq z', \ \tau \neq \tau',$$

and furthermore

$$\sqrt{x^2 + y^2} \neq \sqrt{(x')^2 + (y')^2}$$

(just as on p. 108 of this chapter)
and indeed here
we even have

$$\sqrt{x^2 + y^2 + z^2} \neq \sqrt{(x')^2 + (y')^2 + (z')^2}.$$

This last statement is
MOST STARTLING,
since it says that
the LENGTH of an object
as measured by K
is DIFFERENT from
the length of the SAME object
as measured by K′—
in other words,
Einstein found that
the LENGTH OF AN OBJECT
is NOT an INVARIANT
UNDER THE LORENTZ TRANSFORMATION
(Note 3),
but is more like a SHADOW
(as on pp. 107–108 of this chapter),
and further
the invariant d on p. 109
is NOT the DISTANCE BETWEEN POINTS
(as is the LENGTH of an object)
but is rather
the SPACE-TIME INTERVAL
BETWEEN TWO EVENTS,
which is the
INVARIANT in our
FOUR-DIMENTIONAL SPACE-TIME WORLD,

$$\sqrt{X^2+Y^2+Z^2+T^2} = \sqrt{(X')^2+(Y')^2+(Z')^2+(T')^2}$$

under the
Lorentz Transformation (see p. 147 Note 3).

The reader cannot be expected
to fully appreciate this
without further details,
but what we hope he
CAN appreciate
is the general idea that
two DIFFERENT observers,
call them K and K′,
using DIFFERENT coordinate systems
(different "viewpoints")
and differing also in many other ways
(see p. 110),
may still find AGREEMENT
on the INVARIANTS under
a given transformation
and may thus study the physical universe
WITH EQUAL RIGHT AND EQUAL SUCCESS.

In other words,
in spite of many differences between
various observers,
like K and K′,
they can find SOMETHING to AGREE on.

Neither one
has COMPLETE freedom
to do just ANYTHING he wants—
that is,
there is NO ANARCHY here—
and,
on the other hand,
there is also
NOT COMPLETE REGIMENTATION,
since K and K′ etc.
may each have

his own coordinate system
(his own "viewpoint")
but he is RESTRICTED by
his own set of postulates
as well as by the
METAMATHEMATICS
behind the scenes.*

Thus you see that in
both mathematics and art,
as well as in science,
there is FREEDOM of choice
to a certain extent,
but it is NOT an ALL-OUT FREEDOM
(Anarchy)
and therefore there is no resentment
against certain controls.

And it is this combination of
FREEDOM and CONTROL
that have made
mathematics and science
both CREATIVE and CONSERVATIVE,
as you have seen.
No doubt you will agree that
this COMBINATION of
FREEDOM and CONTROLS
CAN SERVE AS A MODEL
for THINKING about ANYTHING—
whether it be Democracy
or Censorship
or whatever.

* If you wish to know more about
METAMATHEMATICS,
see the very illuminating paper
"Elementary Topics in Mathematical Logic"
by Alonzo Church of Princeton University,
briefly summarized in
Note 2, p. 144.

3. To Sum Up

A. If you are dealing with
any branch of
PURE mathematics,
you must ask,
(1) What are the ELEMENTS?
(2) What are the OPERATIONS?
(3) What are the POSTULATES?

and, of course,
(4) What are the RESTRICTIONS
imposed by
the METAMATHEMATICS?

B. If you are dealing with

any branch of
APPLIED mathematics
(whether applied to physics
or whatever)
you must ask,
in addition to the four questions
mentioned in A above
also:
(5) Have you made
CAREFUL OBSERVATIONS
in the domain in question?

(6) Have you derived the
CONSEQUENCES from
your postulate set,
and CHECKED THEM
against further observations
perhaps discovered later?

And of course if you find that
the "consequences"

do NOT correspond
to the new observations,
you will have to make
CHANGES
in the mathematical system
which you have been using
since it does not correspond to the
"facts."

This is exactly what happened
when new observations were made
by Michelson and Morley in 1887,
which did NOT correspond to
the mathematical system of
Newtonian physics,
in which
POINTS were the ELEMENTS and
TIME was the SAME for
all observers.
And it was EINSTEIN, in 1905,
who suggested that if
EVENTS were taken as ELEMENTS,
and if each observer,
say K and K′,
had his OWN TIME,
t and t′ respectively
Einstein could then develop
a NEW system
(the Special Theory of Relativity)
which not only
more adequately
corresponded to the
actual physical universe
known up to that time,
but,
when he developed it further, in 1915,
into the
General Theory of Relativity,

he was able to
PREDICT observations which
had NOT YET been made,
but which,
when checked,
proved to be CORRECT!

But Einstein knew perfectly well
that his system, too,
would be valid only until
further observations would be made
which would necessitate
further changes in the system.
Or, as he put it,
"ALLES WAS WIR MACHEN IST FALSCH"
(Everything we make is false—
that is, in the long run.)
But this does not happen every minute,
so that physicists now know
perfectly well
that they, even now,
do not know THE truth,
and probably never will,
and yet,
for the time being,
they have a system
that works very well:
they rely on
the INVARIANTS under
certain specific transformations,
without mistaking these for
ETERNAL VERITIES.
And, of course, this means that
each one must make his
measurements ACCURATELY,
must make his
calculations CORRECTLY,
must think LOGICALLY—

all these so that they are
ACCEPTABLE to
the best mathematicians,
the best scientists,
the best logicians,
who are
CHECKING each other's work
in order to prevent
CONTRADICTIONS,
LIES,
or,
as Professor A. Fraenkel
(of the Hebrew University)
pults it:
"WOLVES in SHEEP'S CLOTHING."
And he adds,
speaking of
Cantor's Theory of Transfinites
(though it may as well apply
to many other branches of
mathematics, etc., as well):

We have succeeded in
building a brick wall
around our flock of sheep,
having first thrown out
all KNOWN "wolves in sheep's clothing,"
but we can NEVER be sure that
there are not still other
"wolves in sheep's clothing" present,
which we have NOT YET found out,

and must therefore
ALWAYS be on the watch for them.
And so the task is NEVER DONE.
In other words,
Mathematics and Science,
as well as Art and

ALL human activities,
though they do NOT
profess to be PERFECT,
have certainly met with more success
in solving their problems
when they have followed the pattern
outlined above in A. and B. (p. 114)

In short,
we must take SAM as our MODEL,
in which the S represents
the OBSERVATIONS we make,
the A represents
the INTUITION required
to select a proper postulate set
to fit the observations,
and the M represents
the DEDUCTIONS obtained
from the observations
and this postulate set.

Thus SAM is a better MODEL than
a "scientist,"
who may really be
just a "bottle washer"
instead of a real
SAMite!

V. LIFE, LIBERTY AND THE PURSUIT OF HAPPINESS

1. Behind the Scenes

First of all
may I recall the
BASIC IMPORTANCE of
that principle of METAMATHEMATICS *
which DEMANDS that
a set of postulates for ANY system
be CONSISTENT,
that the postulates
DO NOT CONTRADICT each other
which means of course that
LYING
CANNOT SERVE as an
instrument of thought!
Now is not this statement
usually considered to be
a MORAL principle?
And yet
without it we cannot have
ANY satisfactory mathematical system,
nor ANY satisfactory system of thought—
indeed we cannot even PLAY a GAME properly
with CONTRADICTORY rules!

In a similar way,
I wish to make the point that
there are other important MORAL ideas
BEHIND THE SCENES,
without which there cannot be

* Note 2, p. 144.

ANY MATHEMATICS or SCIENCE.
And therefore, in this sense,
Science is NOT AT ALL AMORAL—
any more than one could have
a fruitful and non-trivial postulate set
in mathematics
which is not subject to
the METAmathematical demand for
CONSISTENCY!
Now, what are some of these OTHER ideas
Behind the Scenes?

2. Life

Let me suggest only a very few—
and if we accept these,
it will be interesting to see
what follows from them.
Let us take a basic postulate here:
LIFE.
For without LIFE
there can be
no living thing—
no flowers,
no animals,
no human race—
also of course
no music, no art,
no science,
no mathematics.

I am not suggesting that we consider here
WHETHER life is worth living,
whether it would make more "sense"
to commit suicide,
whether it is all just
"Sound and fury, signifying nothing."
I am proposing that

LIFE
be taken as a POSTULATE,
and therefore not subject to proof,
just like any other postulate.*
But I propose to MODIFY this
and take more specifically as

POSTULATE I:
The preservation of
LIFE FOR THE HUMAN RACE
is a goal of human effort.
This does not mean that
we are to go about
wantonly killing animals,
but to do this only when
it is necessary to support
HUMAN life—
for food,
for prevention of disease,
vivisection, etc.
Indeed a horse or dog or other animals,
through their friendliness and sincerity,
might actually HELP to sustain
Man's spirit and faith and even his life.
And I interpret this postulate
also to mean that
so-called "sports,"
like bull-fighting,
or "ganging up" on one little fox—
a whole gang of men and women
(and corrupting even horses and dogs
to help!)—
is really a cowardly act,

* Please remember that
a POSTULATE is
Not a "SELF-EVIDENT TRUTH"
but only an ASSUMPTION (See Ch. I)
or an UNPROVED THEOREM.

the "natural" flow—
thus water must be FORCED uphill, etc.—
this force being supplied by
a LIVING WILL.
More generally,
the tendency formulated by the
Second Law—
if not counteracted by other tendencies—
would make the physical world
"run down," "degenerate" to
what is sometimes called "heat death."
Now it seems that
living organisms,
and, in particular, HUMAN organisms,
are capable of
introducing directions of change
that OPPOSE
this degeneration, this death,
this DESTRUCTIVE "DEVIL."
May I therefore suggest as

POSTULATE II:
Each INDIVIDUAL HUMAN BEING
must fight this "degeneration,"
must cling to LIFE as long as possible,
must grow and create—
physically and/or mentally.
And for this we need

POSTULATE III:
We must all have the LIBERTY to
so grow and create,
without of course interfering
with each other's growth,
which suggests

POSTULATE IV:
This Freedom or LIBERTY

so unsportsmanlike that it is amazing
how this activity could ever be called
a "sport."
All this is by way of interpreting
the meaning to be given to
Postulate I:
ACCEPTANCE of LIFE for the HUMAN RACE.
Surely everyone will accept the idea that
is definitely present,
behind the scenes,
in science or mathematics.
But this is not all.
For, I take this postulate to mean also
that we are not to limit it to
only a PART of the human race,
as Hitler did,
because this inevitably leads to WAR,
and in this day of
nuclear weapons
and CBR (chemical, biological, radiological)
weapons,
this would certainly contradict
Postulate I,
would it not?

2. A *System* of MORALITY

And, if we are trying to suggest
a "SYSTEM" of MORALITY,
as a METAMATHEMATICAL background,
it must of course itself be CONSISTENT!
May I say at the very outset that
the "SYSTEM" suggested here
makes no pretense of finality(!),
remembering how difficult it is,
EVEN in MATHEMATICS,
to have a postulate set which is
perfect (see p. 132, Ch. V)!

Nevertheless, one must go on,
one must TRY,
one must do one's BEST,
as in mathematics and sciences.
And so, let us continue, in all humility,
to try to make
what can only at best be regarded as
tentative suggestions,
in the hope that the basic idea—
that there is a MORALITY behind the scenes
in Mathematics and Science—
may prove to be helpful
and may be further
improved and strengthened
as time goes on.

Now,
let us consider a SECOND postulate.
But first may I recall
what physicists call
the two laws of thermodynamics:

A. The Conservation of Energy, namely,
the total energy in a closed system
is a CONSTANT—
it can neither be created nor destroyed,
but only transformed from
one kind of energy to another kind—
thus we can transform
electrical energy into heat.
or heat to mechanical energy, etc.,
without either
adding to or subtracting from
the total energy.
This FIRST LAW of THERMODYNAMICS
HAS SURVIVED EVEN
the great and fundamental changes
made in PHYSICS by
the EINSTEIN Theory of Relativity!

For, when we say that
atomic energy is CREATED,
we mean that
the lost MASS has been transformed
into ENERGY,
since Einstein showed that
MASS and ENERGY are really the same,
as expressed in his famous equation

$$E = mc^2$$

(where E = energy, m = mass, and
c = the velocity of light).
Thus even in the case of atomic energy,
there is no change in
the TOTAL amount of energy,
but merely a transformation
from one kind to another.

B. The SECOND LAW of
THERMODYNAMICS
may be stated:
Whenever there is
a transformation or flow of energy,
it happens always
in a certain "natural" DIRECTION
from a point of higher potential
to one of lower potential,
and never in reverse—
thus water always flows DOWN hill
heat always flows from
a point of higher temperature
to one of lower temperature,
etc., etc.

And the only way
this direction can be reversed
is by FORCING it
in a direction opposite to

must be accompanied by
RESPONSIBILITY,
if it is not to lead to
CONFLICT between
individuals or groups
which would of course
CONTRADICT the other postulates.

All this is of course very DIFFICULT to do,
involving
accepting LIFE without whimpering,
growing without interfering with
the growth of others,
in short
it involves what Goethe called
"cheerful resignation"
("heitere Resignation").

But how can this be done?
It seems clear that we must now add

POSTULATE V:
The PURSUIT of HAPPINESS *
is a goal of human effort.
For without some happiness,
or at least the hope of some happiness
(the "pursuit" of happiness)
it would be impossible
to accept "cheerfully"
the program outlined above.
And such acceptance leads to
a calm, sane performance of our work,
in the spirit in which a mathematician
accepts the postulates of a system

* The reader may be interested to read
the discussion on "happiness" in
another book of this series:
Mits, Wits and Logic,
published by W. W. Norton & Company, New York.

and accepts his creative work
based on these—
accepting even the Great Difficulties
which he encounters
and is determined to conquer.

I repeat that the
"*POSTULATES*" FOR MORALITY
given above
are very tentative and inadequate.
What I am trying to emphasize is:

(1) that there IS a morality
behind the scenes
in mathematics and science,
that they are NOT AMORAL;
(2) that this tacit MORALITY
must be made EXPLICIT
in order to be effective—
just as the idea of METAMATHEMATICS
is becoming tremendously effective
due to its EXPLICIT formulation; *
that, similarly,
the IMPLICIT postulates of MORALITY
behind mathematics and science
(as suggested by the above discussion)
be made EXPLICIT and EFFECTIVE
and thus free these
wonderful human activities from
the false accusation of being
AMORAL.

And I firmly believe that
the results of such a formulation
will re-discover the conclusions
reached by the
great religious leaders and the

* See Note 2, p. 144.

great humanitarians.
But, coming from the
mathematical and scientific camp
it will take a form which will be
more accessible to more people,
who may not have
the great INTUITION needed to
arrive at these conclusions
without such scientific formulation.

May I quote here, in support of the above,
from Einstein's *Out of My Later Years:* *
in speaking of our religious tradition
which has given us
"the highest principles for
our aspirations and judgments";
he then adds:
"If one were to take that goal
out of its religious form,
and look merely at its purely human side,
one might state it perhaps thus:
free and responsible development
of the individual,
so that he may place his powers
freely and gladly in the service of
all mankind."

Do I hear some reader saying
"But how can you say that
Science is NOT AMORAL when
it gives us atomic energy and
DOES NOT CARE whether
we use it for peaceful purposes or
for making bombs?!"
To which I say that

* Published by the
Philosophical Library in 1950,
page 23.

since the bomb DESTROYS LIFE,
therefore the ideal pursued by Science
is INCOMPATIBLE with
the consequences that flow from
the use of the BOMB.*
So it is NOT SCIENCE in Man
but the "Devil" in him
whose objective is DESTRUCTION
that is consistent with the bomb.
Indeed, does not the "Devil" in Man
use even
RELIGION itself for DESTRUCTION?
And yet
you would not say that
RELIGION is AMORAL,
would you?

And
to those psychologists who tell us
that there is a deep desire
for CONFLICT
in all of us,
may I say that
the above program is
in complete agreement,
BUT
it is the "degeneration,"
the "Devil,"
which LIFE MUST FIGHT (see p. 126, Ch. V)
and it is NOT our "grandmothers"
whom we wish to murder,
as some people wrongly interpret
this BASIC urge to FIGHT the "Devil."

* In Hiroshima and Nagasaki:!
and have you heard that
the bombardier who threw
the atomic bombs on
those cities
has since gone insane?!

Thus Man,
being, on the one hand,
part of the physical world
(that is, being partly a "thing")
is governed by
the Second Law (p. 125)
like any other "thing."
BUT on the other hand,
the LIFE in MAN
MUST
according to POSTULATE II
FIGHT this destructive force
within himself.
And this "WAR"
demands our greatest efforts,
so we had better
stop wasting these efforts on
fighting each other
in contradiction to
Postulates III and IV.
Thus,
the above postulate set
inadequate as it probably is in
its present form,
is, nevertheless, I submit
behind the scenes in SCIENCE—
in METASCIENCE—
and is in agreement with
the basic principles given us by
the great religious leaders
and humanitarians,
and is definitely
NOT a POLLYANNISH program!
It is a TOUGH FIGHT,
which each one of us,
as individuals,
is sure to lose!—
for the Second Law will

Surely get us in the end,
by Death—
but which
LIFE itself wins!

If we accept this
"gladly,"
as Einstein says (p. 129)
we have that "happiness" (Postulate V)
without which we could not
"fight the good fight"—
the most interesting and
exhilarating
of all "games"!
How can anyone say that
Science,
with its "METAscience"
behind the scenes,
is AMORAL?!
Is it not rather
the GADGETS themselves,
the BOMBS themselves,
that are AMORAL
(being only "things"),
since they will explode,
indifferently,
to kill millions of people or
to build a new harbor—
but they are not to be identified with
the SPIRIT in man which is
in his SCIENTIFIC URGE! *

3. Can We Do It?

Some appreciation of
the SUBLETY of the demands upon
a Postulate Set,

* The urge to CREATE,
to climb Mt. Everest,

may be seen from the following:
Alfred Tarski, the great Logician,
says in his
"Introduction to Logic,"
pp. 136, 137:
"There are only a few deductive theories known
of which it has been possible to show
that they are
CONSISTENT and COMPLETE."
And further:
"Perhaps the most interesting example of
a consistent and complete theory
is that supplied by
elementary geometry."
And again:
"The situation changes essentially
as soon as one goes over to such sciences as
arithmetic and advanced geometry.
Probably no one
working in these sciences
doubts their consistency;
and yet as has resulted from
the latest methodological investigations,
a strict PROOF of their CONSISTENCY
meets with great difficulties
of a fundamental nature.
The situation in regard to
the problem of COMPLETENESS
is even worse:
it turns out that
arithmetic and advanced geometry
are INCOMPLETE;
for it has been possible
to set up problems of a
purely arithmetical or geometrical character
that can be

---

to examine the Earth, the Seas, and now Space,
and, of course, Human Nature itself!

NEITHER positively NOR negatively decided
within these disciplines."
And worse yet:
"It might be supposed that this fact
is merely an outcome of
the imperfection of the axiom systems
and methods of proof
at our disposal up to date,
and that a suitable modification
(for instance,
an extension of the axiom system)
may in the future, yield complete systems.
Deeper investigations, however,
have shown this conjecture to be erroneous:
NEVER will it be possible to build up
a consistent and complete deductive theory
containing as its theorems
all true sentences of
arithmetic or of advanced geometry."
"These exceedingly important achievements
are due to the contemporary
Austrian logician
K. Gödel."

Is this not reminiscent of
Einstein's famous remark:
"Alles was wir machen ist falsch"
(Everything we make is false)?
as well as Fraenkel's remark on p. 117, Ch. IV.
And yet,
let us not forget that,
even though
Mathematics and Science are
NOT PERFECT,
still they have been
eminently successful,
including of course
Einstein's own work.

Thus we have much to learn from
the methods used in these domains,
even while admitting that
NOTHING that MAN does is PERFECT!

It would seem that
these formidable subtleties
might frighten Man away
entirely
from the pursuit of
science and mathematics!
But the great research men
are made of "sterner stuff,"
and
whereas they continue to dig
into these subtle foundations,
they, at the same time,
continue to build the
superstructure—
and with tremendous success!

VI. THE UPSHOT!

1. Postulational thinking

Thus we have seen that
whereas mathematics and science
are NOT PERFECT (see pp. 133–135, Ch. V),
and, indeed,
the greater the AREA of our knowledge,
the greater also becomes
the CIRCUMFERENCE in contact with
the surrounding UNKNOWN,*
STILL
with the use of
POSTULATIONAL THINKING
as described in previous chapters,
Man has been able
to make up
a large variety of systems
which can be applied to
the successful solution of
a tremendous variety of problems,
as you have seen.
This postulational method of thinking
has even been successfully applied
to the study of the universe itself
(Chapter IV),
allowing different observers,
K and K′,
each to use his own
coordinate system,
his own "viewpoint"
and yet find the

* P. 9 of the "Prologue and Epilogue."

INVARIANTS
that they can agree on.

2. Morality

As we have seen (Ch. V)
Postulational Thinking
is even helpful in
considering a SYSTEM FOR MORALITY.
And what is the BASIC postulate
we suggested here?—
LIFE for the HUMAN RACE
(as interpreted on pp. 122–123, Ch. V).
This suggests other postulates
(see Ch. V),
all of which can be
acceptable
to both K and K′,
even though they have
DIFFERENT coordinate systems.

3. Some Invariants and Some Differences

(1) Invariants: LIFE—which demands
    (a) Sufficient and proper food;
    (b) Good Health;
    (c) Education—both mental and physical;
    (d) NO VIOLENCE!
    (a real scientist does NOT go
    into his laboratory with an AXE
    with which to DESTROY his apparatus,
    but rather with a well-developed BRAIN,
    and lots of PATIENCE
    with which to CREATE new things
    which will be BENEFICIAL to the
    HUMAN RACE).
    This of course implies PEACE,
    and better still

Peace, friendship, and work.

(e) FRIENDSHIP between K and K′!
(f) Humility—remembering that
he will NEVER know THE "truth"
(see Ch. V, pp. 133–135) etc., etc.
(g) And all this of course
requires a great deal of
HARD WORK.

(2) Differences which will
NOT PREVENT both K and K′
from studying the universe
WITH EQUAL RIGHT and EQUAL
SUCCESS—
which is certainly
the clearest concept of
what DEMOCRACY is:
(a) Different coordinate systems (Ch. IV)
(b) Differences in color of skin!
(c) Different languages—or
other means of communication.

And please do not consider this program
as an unattainable "Utopia,"
for it really WORKS in
Science and Mathematics,
as we have seen,
and can also work in
other domains,
if we would only
put our BEST EFFORTS into it,
instead of
fighting WARS—
(HOT or COLD)
or even PREPARING for wars—
HATING other people,
etc., etc.

If you really consider the matter
CAREFULLY,

The last warrior.

you will agree that
we MUST overhaul
most of our so-called "thinking" *—
and

if you have read this book—
not just "skimmed" through it—
you will surely admit that
we have a wonderful *model* in
Science and Mathematics,
which should be followed also in other domains—
even MUSIC, ART etc.—
which have their own
(1) ELEMENTS
(2) OPERATIONS
(3) POSTULATES.

These give DIRECTION to our efforts,
instead of
just running around from
one cocktail party to another,
and worrying about our "status,"
and wasting our
wonderful potentialities.

* See Note 1.

VII. IMPORTANT NOTES

Note 1

EINSTEIN'S TELEGRAM TO THE PEOPLE:

Our world faces crisis
as yet UNPERCEIVED by those
possessing power to make
great decisions for good or evil.
Unleashed power of atom
has CHANGED EVERYTHING
EXCEPT our modes of THINKING
AND WE THUS DRIFT TOWARD
UNPARALLELED CATASTROPHE.
We scientists who released this immense power
have overwhelming responsibility
in this world-wide life-and-death struggle
to harness atom
for BENEFIT of MANKIND
and NOT for humanity's destruction.
Bethe, Condon, Szilard, Urey and
Federation of American Scientists
join me in this appeal
and beg you to support
our efforts to bring realization to America
that mankind's destiny is being decided
TODAY
NOW
THIS MOMENT!

And this is truer than ever NOW, in 1961.

And now the atom.

Note 2

METAMATHEMATICS:

Let me refer you to
"Elementary Topics in Mathematical Logic"
by Professor Alonzo Church of
Princeton University,
in which,
in developing SET THEORY,
he explicitly states that
he will introduce
TWO KINDS of postulates:

(1) those WITHIN the system
which are entirely expressible
in the notation of set theory,
and

(2) those which are
NOT IN
but ABOUT
set theory,
which "require for their expression
the use of English * words
in addition to
the notation of set theory."

This second kind of postulates
serves the following functions:

they EITHER

(a) give permission to ASSERT
any sentence of a certain kind or form,
in which case they are called
SCHEMATA, or

* Or, of course, any other native language.

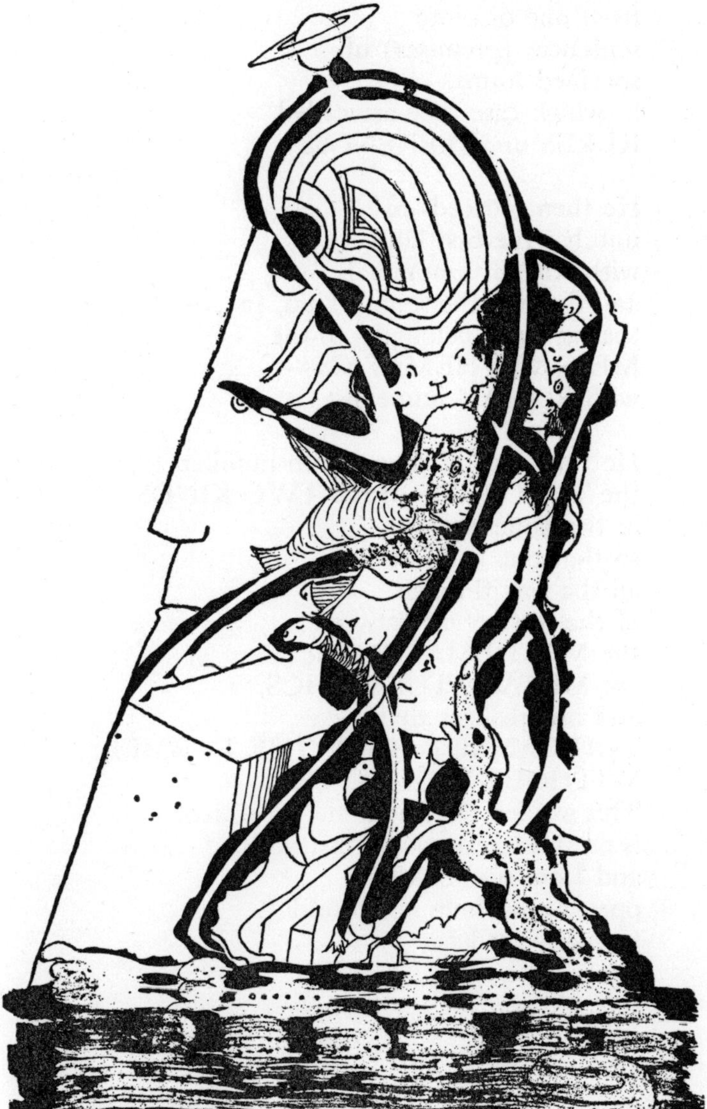

(b) give permission to INFER
a sentence of a certain form
from one or more
sentences (premises) of
specified forms,
in which case they are called
RULES of INFERENCE.

He then proceeds to
number the first kind, (1),
with ARABIC numerals,
and those of the second kind, (2),—
that is, those belonging to the
METAMATHEMATICS—
with ROMAN numerals.

He then further proceeds to number
the THEOREMS of the TWO KINDS
in the same way,
so that one is aware
all the way through
of the distinction between
the MATHEMATICS and
the METAMATHEMATICS,
and one realizes that
YOU CANNOT HAVE THE FORMER
WITHOUT THE LATTER.
This explicit statement and treatment
is most interesting and instructive
and I am sure it will
open up a whole new world
to many readers.

Note 3

THE LORENTZ TRANSFORMATION:

$$\left.\begin{aligned} x' &= \beta(x - vt) \\ y' &= y \\ z' &= z \\ t' &= \beta(t - vx/c^2). \end{aligned}\right\} *$$

See *The Einstein Theory of Relativity*
by Lillian R. Lieber
with drawings by
Hugh Gray Lieber,
published by
Holt, Rinehart, and Winston.
An invariant here is

$$\sqrt{x^2 + y^2 + z^2 + \tau^2}$$

as was stated on p. 109 of Ch. IV.

* Where $\beta = c/\sqrt{c^2 - v^2}$

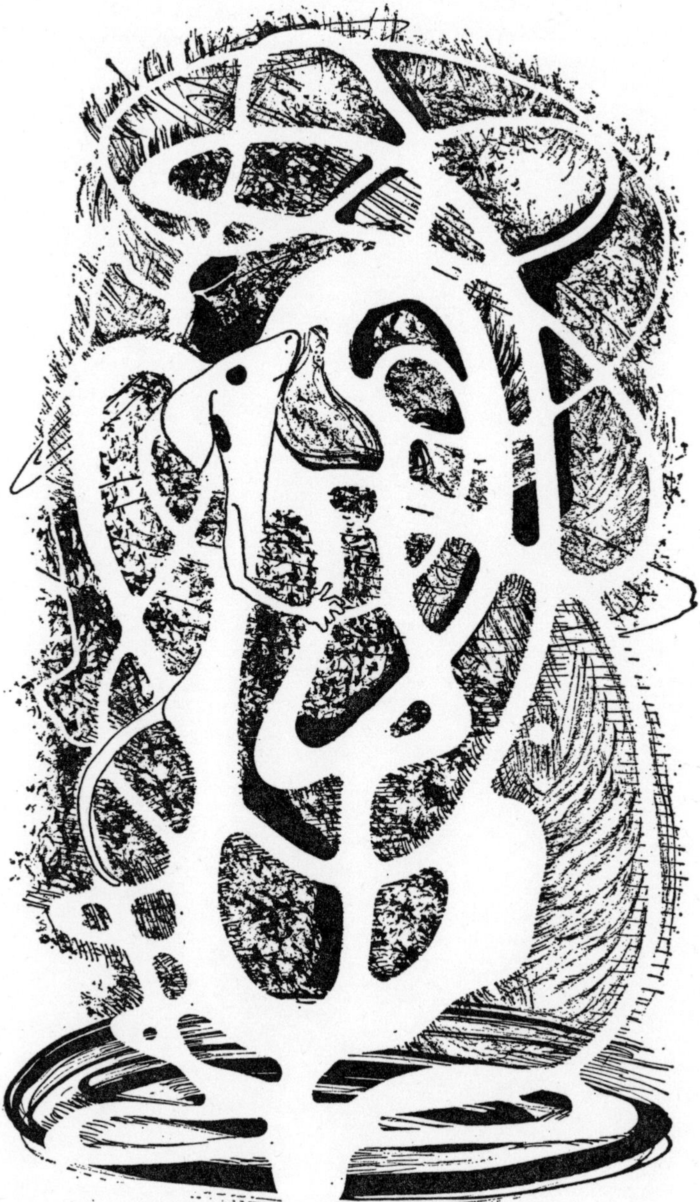

First earth-artist astronaut sketching on Mars.